The Rise of MXenes:
A New Frontier in
Two-Dimensional Materials

Shah

TABLE OF CONTENTS

Chapter 1- Introduction

1.1 The Rise of MXenes

For the past decade, two-dimensional materials have been widely studied since the discovery of single-layer graphene. The unique physical and chemical properties prompt the huge research interest both on well-known two-dimensional materials, such as reduced graphene oxide and metal dichalcogenides, and the discovery of new 2D materials. In 2011, a series of transition metal carbides, carbonitrides and nitrides known as MXenes were discovered at Drexel University[1]. In a 2D flake of $M_{n+1}X_nT_x$ MXene, n layers of carbon or nitrogen (X) interleave with n+1 (n=1-3) layers of early transition metals (M). The Tx stands for the surface terminations boned to the outer M layers, such as hydroxyl, oxygen, fluorine, and/or chlorine[2]. With wide chemical and structural variety, MXenes have shown significant promise in energy storage applications[3-5] and great potential in catalysis[6-7], electronics[8-9], electromagnetic[10-11], biomedical[12-13], and other applications[14-15]. In this book, we demonstrate the excellent versatility of MXenes by showing for the first time the ability to grow ferroelectric crystals and reinforce nanocomposites.

1.1.1   MAX Phase

MXenes are made by selective etching of certain atomic layers from their layered precursors, such as MAX phases. MAX phases are a very large family of ternary carbides and nitrides with more than 70 reported so far, in addition to numerous solid solutions and ordered double transition metal carbides or nitrides $(M_{n+1}X_n)$[16-19]. The metallic M-A bonds are more chemically active than the stronger M-X bonds, which makes selective etching of A-element layers possible. Figure 1.1 demonstrate the composites of MAX phases. The elements with blue background (M) represent the early transition metal reported in MAX phases, including Ti, V, Cr,

Y, Nb, Mo, Hf, Ta, and W. The group A elements are marked with red background, such as Al, Si, P, and S, which can potentially be selectively etched to make MXenes. From several different group A elements, so far only Al has been successfully etched from MAX phase to form MXenes. X elements include carbon and nitrogen. In this book, Ti_3AlC_2 and Nb_2AlC MAX phases are used to synthesize $Ti_3C_2T_x$ and Nb_2CT_x MXene, respectively.

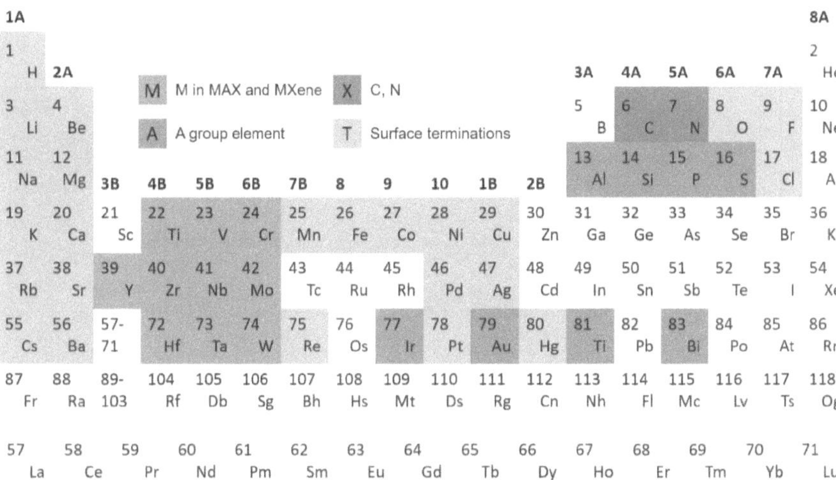

Figure 1.1 Periodic tables showing compositions of MAX phases.

1.1.2 MXenes

In order to synthesize two-dimensional MXenes, aqueous fluoride-containing acidic solutions, either pristine hydrofluoric acid (HF)[20] or in-situ formatting HF through the reaction of hydrochloric acid (HCl) and lithium fluoride (LiF)[21], are always used to selectively etch the A atomic layers from MAX phases. Figure 1.2 shows both experimentally (marked in blue) and theoretically (marked in gray) explored compositions of MXenes to date, which present three types of formula, M_2X, M_3X_2, and M_4X_3. There are two types of M element, mono-M (Ti_2C, Zr_3C_2, and V_4C_3) and double-M. Usaully, double-M MXenes include solid solution double-M (($Ti,V)_2C$, ($Ti,Nb)_2C$) MXenes and ordered double-M MXenes (($Mo_2Ti)C_2$, ($V_2Ti)C_2$, ($Ti_2Ta_2)C_3$).

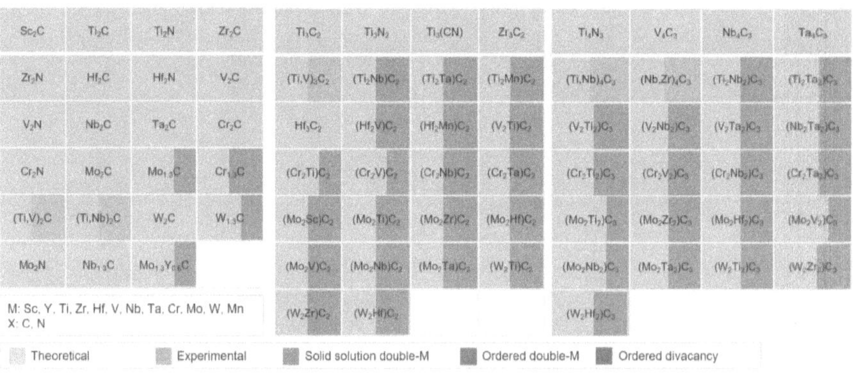

Figure 1.2 MXene compositions reported to date.

1.1.3 Key Properties of MXenes

2D MXenes have been proved to possess quite a few key properties, such as high electrical conductivity[22], hydrophilic surface[23], high mechanical stiffness[24], printable[25], photothermal[26], and large electrochemical activity[27], etc (Figure 1.3). The metallic electrical conductivity of $Ti_3C_2T_x$ MXene can reach up to 20,000 S/cm via Four Probes resistivity test system. Thermal conductivity (κ) of $Ti_3C_2T_x$ MXene was calculated using equation $\kappa = D\ Cp\ \rho$, where D is the thermal diffusivity, Cp is specific heat capacity and ρ is the density calculated by dividing the mass by the volume. MXenes are hydrophilic due to the oxide/hydroxide-like polar surface, which makes them environmentally friendly and printable from aqueous solutions. In addition, $Ti_3C_2T_x$

MXene has strong absorption in the UV-visible-NIR range, which makes it a good candidate for photothermal therapy, photo-thermo-mechanical actuator, plasmon-assisted catalysis.

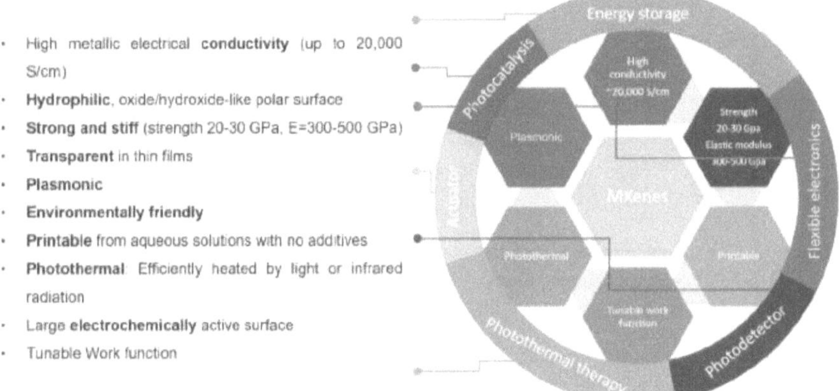

Figure 1.3 Key properties and corresponding applications of MXenes

1.2 Experimental Methods to Synthesize MXenes

1.2.1 Sintering

Nb_2AlC powders were synthesized by mixing powders of niobium (Alfa Aesar, 325 mesh, 99.8%), aluminum (Sigma Aldrich, 325 mesh, 99.5%), and graphite (Sigma Aldrich, 300 mesh, 99.99%) in a molar ratio of 2.0:1.1:1.0, respectively. The powders were further ball-milled at 100 rotations per minute (rpm) for 18 hours. The mixture was pressed into a cube and then heated in a tube furnace to 1600 °C at a heating rate of 5 °C·min-1 under a flow of Ar. The powder was then soaked at that temperature for four hours. After cooling the furnace to room temperature, lightly sintered blocks were obtained. The powders were then sieved, and only ≤ 45-μm powders were used for further experiments.

1.2.2   Etching and Delamination

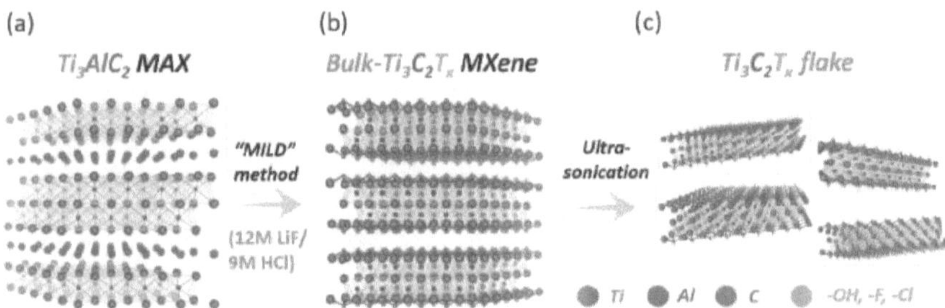

Figure 1.4 Schematic illustration of the structure of the (a) Ti3AlC2 MAX phase, (b) bulk-Ti3C2Tx MXene after etching and delamination, and (c) MXene two dimensional (2D) flakes of different sizes obtained by ultrasonication.

Synthesis of Ti_3C_2 MXene: All chemicals were used as received without further purification. Layered ternary carbide Ti_3AlC_2 (MAX phase) powder was commercially procured (Carbon-Ukraine ltd. particle size < 40 μm). $Ti_3C_2T_x$ MXene was synthesized following the minimally intensive layer delamination (MILD) method, which entailed selective etching of aluminum from Ti_3AlC_2 using in-situ HF-forming etchant as previously reported in extensive detail [28]. The etching solution was prepared by adding 1 g lithium fluoride (LiF, Alfa Aesar, 98+%) to 20 ml 9 M hydrochloric acid (HCl, Fisher, technical grade, 35-38%) followed by stirring for five minutes. Then, 1 g of Ti_3AlC_2 powder was slowly added to the MILD etchant at 35 °C and stirred for 24 hours. The acidic suspension was washed with deionized (DI) water until pH > 6 *via* centrifugation at 3,500 rpm (five minutes per cycle), with decanting of the supernatant after each cycle. Around pH ≥6, a stable dark green supernatant of $Ti_3C_2T_x$ was observed and then collected after five minutes of centrifugation at 3,500 rpm. The concentration of the $Ti_3C_2T_x$ solution was measured by filtering specific amounts of colloidal solution through a polypropylene filter (3,501 coated PP, Celgard LLC, Charlotte, NC), followed by overnight drying under a vacuum at 70 °C.

Synthesis of Nb₂C MXene: One gram of the obtained Nb_2AlC MAX powders was immersed in 10 ml of 48 % concentrated HF solution and stirred for 48 hours at 55 °C. After the HF treatment, the resulting suspension was washed with deionized water and centrifuged (4,500 rpm for three minutes) to separate the settled powders from the supernatant. The washing process was repeated until the pH of the supernatant increased to ~6. The settled powders were then removed from vials with ethanol and dried at 80 °C in the vacuum oven.

1.3 Objective of Research

We wanted to capitalize on the 2-D nature of MXenes by using them as precursors for the synthesis of 2D functional material. Since MXenes are easily intercalated with monovalent cations K, Na, Li due to their expanded d-spacing after etching, we tried to develop new synthesis processes of texture functional ferroelectric, piezoelectric, and photoluminescent crystals using MXenes as precursors. In addition, we would like to utilize 2D nature, high electrical conductivity, and photothermal properties of MXenes to fabricate MXene/polymer composites for high energy density capacitor and energy harvesting applications.

1.4 Multifunctional MXene-derived Materials

1.4.1 MXene-derived Photoluminescent/Ferroelectric Crystals

MXene-derived high aspect ratio potassium niobate ($KNbO_3$) single crystals are successfully synthesized using two-dimensional Nb₂C MXene and KOH as the potassium and niobium source, respectively. The well-defined butterfly loops of the PFM amplitude signals and the distinct 180-degree switching of the phase signals further corroborate the presence of robust ferroelectricity in M-$KNbO_3$ crystals.

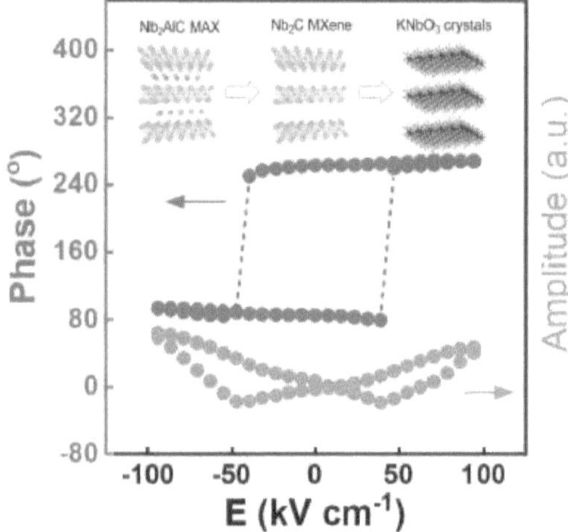

Figure 1.5 Nb$_2$CT$_x$ MXene-derived KNbO$_3$ ferroelectric crystals

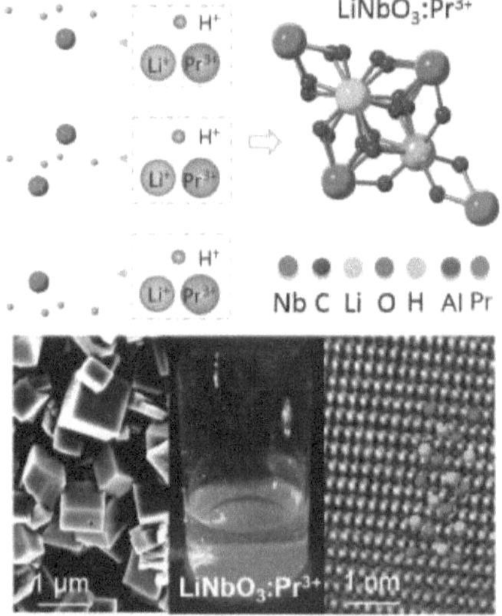

Figure 1.6 Photoluminescent LiNbO$_3$:Pr^{3+} crystals derived from Nb$_2$CT$_x$ MXene

MXene-derived high aspect- ratio lithium niobate (LiNbO3) single crystals are successfully synthesized using two-dimensional Nb2C MXene and LiOH as the lithium and niobium sources, respectively. In addition, for the first time, trivalent rare-earth ions in the M-LN crystals are doped using a different synthesizing process, and provide photoluminescent properties to those doped LiNbO$_3$ crystals.

1.4.2 MXene/polymer Percolative Composites

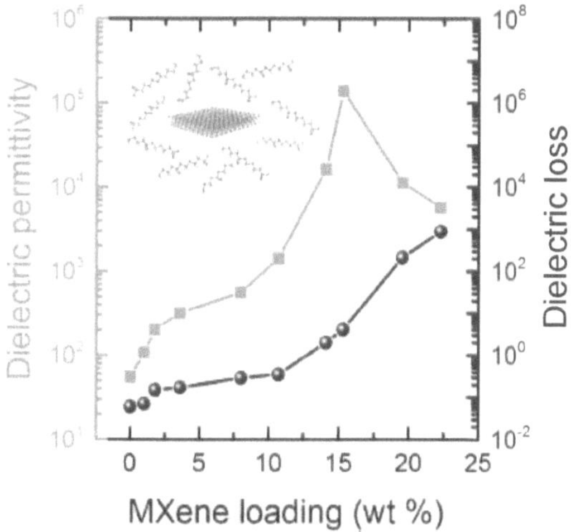

Figure 1.7 Dielectric enhancement in Ti3C2Tx MXene/P(VDF-TrFE-CFE) composites

MXene-containing insulating polymers evidence a large enhancement in dielectric constant. The ratio of permittivity to loss factor for MXene-percolated composites is superior to that of all previously reported fillers in the same polymer. The dielectric constant enhancement is explained on the basis of the microscopic dipole model.

1.4.3 MXene/in-plane Aligned PVDF Actuator

We have demonstrated the solar tracking system based on the photo-thermo-mechanical (PTM) deformation of plasmonic $Ti_3C_2T_x$ MXene/3D aligned PVDF actuator. The sensitivity of the PTM actuator to photo stimulation is due to the marked enhancement of the surface plasmon resonance of $Ti_3C_2T_x$ MXene. In this work, three types of tracking options were simulated, where uniaxial E-W options improved the overall energy intensity of solar module by over 30 % compared to the optimized tilting-control option. Our results clearly show an improvement in PV efficiency, PV power intensity, and energy intensity based on our MXene/PVDF PTM actuator. Our design provides a new way to achieve solar tracking function using photo-thermal-mechanical property of two dimensional materials.

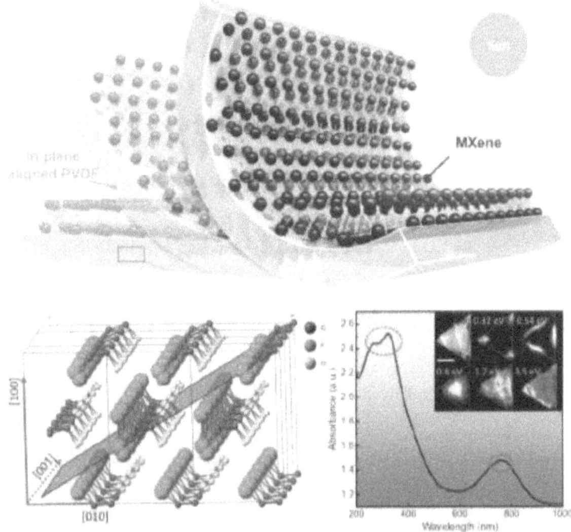

Figure 1.8 Schematic illustration of the photothermal-thermomechanical (PT-TM) actuator based on MXene- aligned PVDF.

REFERENCES

[1]   M. Naguib, M. Kurtoglu, V. Presser, J. Lu, J. Niu, M. Heon, L. Hultman, Y. Gogotsi, M. W. Barsoum. *Adv. Mater.* **2011**, *23*, 4248–4253.

[2]   B. Anasori, M. R. Lukatskaya, Y. Gogotsi. *Nat. Rev. Mater.* **2017**, *2*, 16098.

[3]   M. Naguib, J. Come, B. Dyatkin, V. Presser, P. L. Taberna, P. Simon, M. W. Barsoum, Y. Gogotsi. *Electrochem. commun.* **2012**, *16*, 61.

[4]   O. Mashtalir, M. Naguib, V. N. Mochalin, Y. Dall'Agnese, M. Heon, M. W. Barsoum, Y. Gogotsi, *Nat. Commun.* **2013**, *4*, 1.

[5]   M. Naguib, J. Halim, J. Lu, K. M. Cook, L. Hultman, Y. Gogotsi, M. W. Barsoum, *J. Am. Chem. Soc.* **2013**, *135*, 15966.

[6]   Z. W. Seh, K. D. Fredrickson, B. Anasori, J. Kibsgaard, A. L. Strickler, M. R. Lukatskaya, Y. Gogotsi, T. F. Jaramillo, A. Vojvodic. *ACS Energy Lett.* **2016**, *1*, 58.

[7]   L. Xiu, Z. Wang, M. Yu, X. Wu, J. Qiu, *ACS Nano* **2018**, *12*, 8017.

[8]   C. Zhang, B. Anasori, A. Seral-Ascaso, S.-H. Park, N. McEvoy, A. Shmeliov, G. S. Duesberg, J. N. Coleman, Y. Gogotsi, V. Nicolosi. *Adv. Mater.* **2017**, *29*, 1702678.

[9]   Z. Wang, H. Kim, H. N. Alshareef. *Adv. Mater.* **2018**, *30*, 1706656.

[10]   F. Shahzad, M. Alhabeb, C. B. Hatter, B. Anasori, S. M. Hong, C. M. Koo, Y. Gogotsi. *Science* **2016**, *353*, 1137-1140.

[11]   R. Bian, G. He, W. Zhi, S. Xiang, T. Wang, D. Cai. *J. Mater. Chem. C* **2019**, *7*, 474.

[12]   M. Soleymaniha, M.-A. Shahbazi, A. R. Rafieerad, A. Maleki, A. Amiri. *Adv. Healthcare Mater.* **2019**, *8*, 1801137.

[13]   K. Huang, Z. Li, J. Lin, G. Han, P. Huang. *Chem. Soc. Rev.* **2018**, *47*, 5109.

[14]   K. Hantanasirisakul, Y. Gogotsi. *Adv. Mater.* **2018**, *30*, 1804779.

[15]   S. J. Kim, H.-J. Koh, C. E. Ren, O. Kwon, K. Maleski, S.-Y. Cho, B. Anasori, C.-K. Kim, Y.-K. Choi, J. Kim, Y. Gogotsi, H.-T. Jung. *ACS Nano* **2018**, *12*, 986–993.

[16]   M. W. Barsoum, MAX Phases: Properties of Machinable Ternary Carbides and Nitrides (Wiley, 2013).

[17]   M. W. Barsoum, M. Radovic. *Annu. Rev. Mater. Res.* **2011**, *41*, 195–227.

[18]   P. Eklund, M. Beckers, U. Jansson, H. Högberg, L. Hultman. *Thin Solid Films* **2010**, *518*, 1851–1878.

[19]   B. Anasori, M. Dahlqvist, J. Halim, E. J. Moon, J. Lu, B. C. Hosler, E. N. Caspi, S. J. May, L. Hultman, P. Eklund, J. Rosén, M. W. Barsoum. *J. Appl. Phys.* **2015**, *118*, 094304.

[20]   M. Naguib, V. N. Mochalin, M. W. Barsoum, Y. Gogotsi. *Adv. Mater.* **2014**, *26*, 992–1004.

[21]   M. Ghidiu, M. R. Lukatskaya, M.-Q. Zhao, Y. Gogotsi, M. W. Barsoum. *Nature* **2014**, *516*, 78–81.

[22]   Z. Guo, L. Gao, Z. Xu, S. Teo, C. Zhang, Y. Kamata, S. Hayase, T. Ma. *Small* **2018**, *14*, 1802738.

[23]   J. Zhao, Y. Yang, C. Yang, Y. Tian, Y. Han, J. Liu, X. Yin, W. Que. *J. Mater. Chem. A* **2018**, *6*, 16196.

[24]   A. Lipatov, H. Lu, M. Alhabeb, B. Anasori, A. Gruverman, Y. Gogotsi, A. Sinitskii. *Sci. Adv.* **2018**, *4*, eaat0491.

[25]   C. J. Zhang, L. McKeon, M. P. Kremer, S.-H. Park, O. Ronan, A. S. Ascaso, S. Barwich, C. Ó. Coileáin, N. McEvoy, H. C. Nerl, B. Anasori, J. N. Coleman, Y. Gogotsi, V. Nicolosi. *Nat. Commun.* **2019**, *10*, 1795.

[26]   X. Fan, Y. Ding, Y. Liu, J. Liang, Y. Chen. *ACS Nano* **2019**, *13*, 8124−8134.

[27]   Y. Yoon, A. P. Tiwari, M. Lee, M. Choi, W. Song, J. Im, T. Zyung, H.-K. Jung, S. S. Lee, S. Jeon, K.-S. An. *J. Mater. Chem. A* **2018**, *6*, 20869.

[28]   C. J. Zhang, S. Pinilla, N. McEvoy, C. P. Cullen, B. Anasori, E. Long, S.-H. Park, A. S. Ascaso, A. Shmeliov, D. Krishnan, C. Morant, X. Liu, G. S. Duesberg, Y. Gogotsi, V. Nicolosi. *Chem. Mater.* **2017**, *29*, 4848-4856.

Chapter 2- Nb$_2$CT$_x$ MXene-derived KNbO$_3$ Ferroelectric Crystals

This study demonstrates the first synthesis of MXene derived ferroelectric crystals. Specifically, we have successfully synthesized high aspect ratio potassium niobate (KNbO$_3$) ferroelectric crystals using two-dimensional Nb$_2$C MXene and KOH as the potassium and niobium source, respectively. Material analysis confirms that a KNbO$_3$ orthorhombic phase with Amm2 symmetry was obtained. Additionally, ferroelectricity in KNbO$_3$ was confirmed using standard ferroelectric, dielectric, and piezoresponse force microscopy (PFM) measurements. The KNbO$_3$ crystals exhibit a saturated polarization of ~ 21 µC/cm^2, a remnant polarization of ~17 µC/cm^2, and a coercive field of ~50 kV/cm. Our discovery illustrates that the two-dimensional nature of MXenes can be exploited to grow ferroelectric crystals.

MXenes have immense potential in many applications, such as water purification[1], reinforcement in nanocomposites[2], gas- and biosensors[3-4], and photothermal therapy[5]. However, researchers have not sufficiently explored the potential for MXenes in electronic applications. The present study reveals the excellent versatility of MXenes by demonstrating that their two-dimensional character can be exploited to grow ferroelectric crystals.To determine the feasibility of deriving ferroelectrics from MXenes, we chose to synthesize KNbO$_3$, which has several interesting properties and potential applications. Such applications include ferroelectric devices[6], optical waveguides[7], holographic storage systems[8], biocompatible transducer[9], and other electronic devices[10-13]. At room temperature, KNbO$_3$ has an orthorhombic crystal structure that lacks a center of symmetry, hence exhibits ferroelectricity at ambient temperatures. Moreover, KNbO$_3$ is lead free, which makes it an environmentally friendly alternative to lead-based ferroelectrics[9]. The massive potential of KNbO$_3$ has stimulated research on the synthesis of powder, nanorods, nanowires, and films by solid-state reaction[14], a sol-gel process[15], hydrothermal treatment[16-18], and other means[19]. However, no work explored the synthesis of ferroelectric $KNbO_3$ crystals with a large aspect ratio, which could significantly prompt the development of nanoelectromechanical systems[17]. Furthermore, $KNbO_3$ crystals that possess high-aspect-ratio structures could be suitable for utilization in integrated photonics, given that increasing the crystalline size is essential to maximize the electro-optic response[20]. Compared to polycrystalline perovskite materials, high-aspect-ratio crystalline perovskite thin films or crystals can dramatically reduce charge recombination, which would improve efficiency of perovskite oxides in some applications[21-22].

This work details the first use of two-dimensional Nb_2C MXene and a hydrothermal method to synthesize $KNbO_3$ crystals that possess robust ferroelectric properties. Thin-layered $KNbO_3$ crystals were fabricated by selective etching of the Nb_2AlC MAX phase, followed by simultaneous oxidation and alkalization processes of the HF-etched Nb_2C MXene phase. The orthorhombic crystal structure of $KNbO_3$ was confirmed by X-ray diffraction (XRD) and transmission electron microscopy (TEM). A ferroelectric tester was used to evaluate the existence of switching polarization in the $KNbO_3$ crystals, and well-defined piezoresponse force microscopy (PFM) amplitude butterfly loops and 180-degree phase switching further corroborated the robust ferroelectricity.

2.1 Hydrothermal Synthesis

Figure 2.1 shows a schematic of the synthesis process used to derive $KNbO_3$ crystals from Nb_2C MXene. The methods section describes the solid-liquid reaction that was used to prepare the parent Nb_2AlC MAX phase. After this reaction, Nb_2C MXene phase was synthesized by selectively etching the Al layer from the pristine Nb_2AlC phase using HF acid according to previously established protocols[23]. Finally, KNbO$_3$ crystals were fabricated via a hydrothermal method which consisted of oxidation and alkalization of Nb$_2$C MXene in a mixed solution of potassium hydroxide (KOH) solution and sodium dodecyl sulfate (SDS) surfactant (see methods section for details).

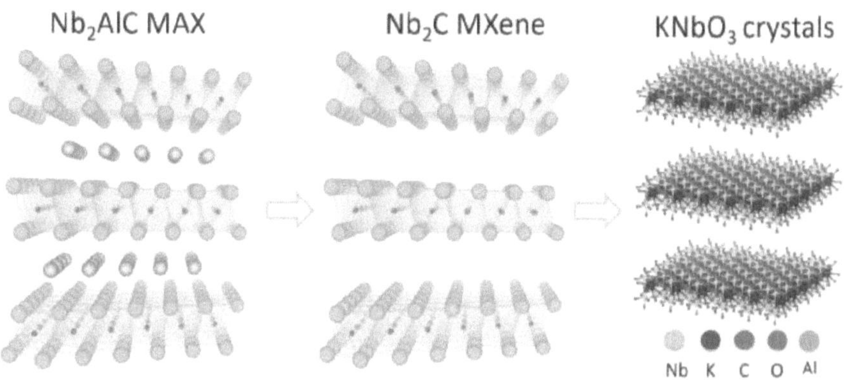

Figure 2.1 Crystal structures of the initial Nb$_2$AlC MAX phase, Nb$_2$C MXene after etching out Al, and MXene-derived KNbO$_3$ ferroelectric crystals (M-KNbO$_3$).

Figure 2.2b depicts a scanning electron microscopy (SEM) of the accordion-like nanostructure, which indicates the successful synthesis of Nb$_2$C MXene from the densely packed Nb$_2$AlC MAX phase (Figure 2.2a), which was additionally confirmed by the XRD patterns. Figure 2.2c illustrates that the (002) peak of Nb$_2$C MXene is significantly shifted to the lower 2θ position, which signals expanded interlayer spacing. Previous work has also found the presence of NbC impurity. The high- resolution X-ray photoelectron spectroscopy (XPS) analysis of Nb$_2$C MXene is shown in Figure 2.2d-f. Figure 5.2d illustrates that the Nb 3d peak can be fitted with six symmetric peaks at 203.8, 204.4, 206.5, 207.0, 207.8 and 209.8 eV, which correspond to Nb-C 3d$_{5/2}$, NbC$_x$O$_y$ 3d$_{5/2}$, Nb-C 3d$_{3/2}$, Nb(V) 3d$_{5/2}$, NbC$_x$O$_y$ 3d$_{3/2}$, and Nb(V) 3d$_{3/2}$[24], respectively. The oxygen signal mainly originates from the intercalated water, surface oxidation and surface functional groups.

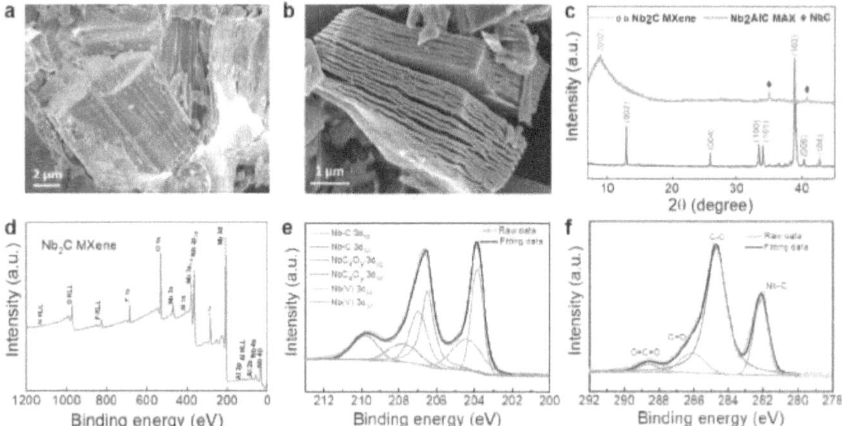

Figure 2.2 Scanning electron microscopy image of (a) Nb₂AlC MAX phase and (b) Nb₂CTₓ MXene phase. (c) XRD pattern of the Nb₂AlC MAX phase and the Nb₂C MXene. The XPS spectrum of (d) Nb₂C MXene, (e) Nb 3d, and (f) C 1s.

The SEM image in Figure 2.3a displays the morphology and size of optimized MXene-derived KNbO₃ (M-KNbO₃) crystals (obtained using Nb₂C MXene, KOH as precursors and SDS as surfactant at 48 hours hydrothermal reaction time), which have an average width of 15 μm and average length of 30 μm. The atomic force microscopy (AFM) image in Figure 5.3b depicts the surface morphology of a 10-nm-thick M-KNbO₃ crystals. The average thickness of M-KNbO₃ crystals is approximately 200 nm (Figure 2.4). Furthermore, scanning transmission electron microscopy (STEM) elemental mapping reflects a homogeneous elemental distribution of K, Nb, and O in the M-KNbO₃ crystal (Figure 2.3c). In addition, the XRD shown in Figure 2.3d verifies the orthorhombic structure of the M-KNbO₃ phase. Figure 2.3e shows the Raman spectrum of M-KNbO₃ crystals, which exhibits strong peaks at 607 cm⁻¹ and 827 cm⁻¹ that can be assigned to the

A_1-ω_{TO4} mode and A_1-ω_{LO4} mode, respectively. The additional peaks at 187cm^{-1}, 190cm^{-1}, 283cm^{-1}, 290cm^{-1}, and 511 cm^{-1} can be respectively assigned to the vibrational modes B_1-ω_{TO1}, A_1-ω_{TO1}, A_2, A_1-ω_{TO2}, and B_2-ω_{TO3}[25]. Figure 2.3f provides the ultraviolet (UV)-visible absorption spectrum of M-KNbO$_3$ crystals dispersed in ethanol. Strong absorption clearly occurs at a wavelength of about 305 nm, and there are no other obvious absorption bands in the whole wavelength range (200~800 nm). The band gap of M-KNbO$_3$ crystals is estimated to be 4.06 eV from the onset of absorption, which is higher than that of KNbO$_3$ cubes (~3.14 eV)[26]. The combined action of size and surface effect result in a blue shift of optical absorption that leads to larger energy gap. The inset in Figure 2.3f displays the photoluminescence (PL) spectrum of the M-KNbO$_3$ crystals, which exhibit a visible light emission peak at 545 nm (2.27 eV). Since the bandgap of M-KNbO$_3$ crystals is approximately 4.06 eV, we can conclude that the light emissions that were observed from 500 nm to 600 nm are not caused by a direct electron transition between the valence and conduction bands. Researchers have found similar results in KNbO$_3$ nanowires and LiNbO$_3$ crystals[27-28]. The luminescence of LiNbO$_3$ can reportedly be ascribed to the recombination of electrons captured by Nb^{5+} ions and holes captured by oxygen, and both site niobium (occupying the position of Nb) and antisite niobium (occupying the position of Li) are responsible for the main emission in LiNbO$_3$. It is well-known that thinner crystals correspond to higher surface-to-volume ratio and higher concentration of surface oxygen vacancies. Therefore, it is likely that the emission band centered at a wavelength around 545 nm of M-KNbO$_3$ crystals can be derived from oxygen vacancies, the site niobium, and antisite niobium.

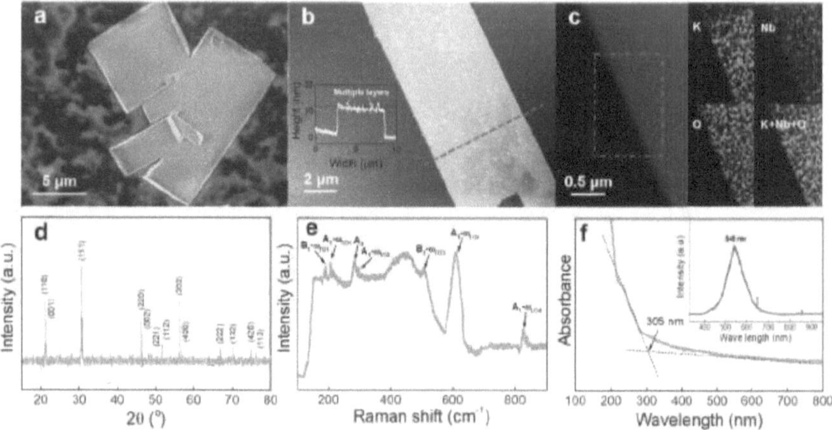

Figure 2.3 (a) Scanning electron microscopy image and (b) atomic force microscopy image of M-KNbO₃ crystal. (c) Scanning transmission electron microscopy image of KNbO₃ crystal, which indicates the elemental distribution in the M-KNbO₃ crystal. (d) XRD pattern and (e) Raman spectrum of M-KNbO₃ crystals. (f) The UV-vis absorbance of M-KNbO₃ crystals in alcohol. Inset: Photoluminescence spectrum of M-KNbO₃ crystals.

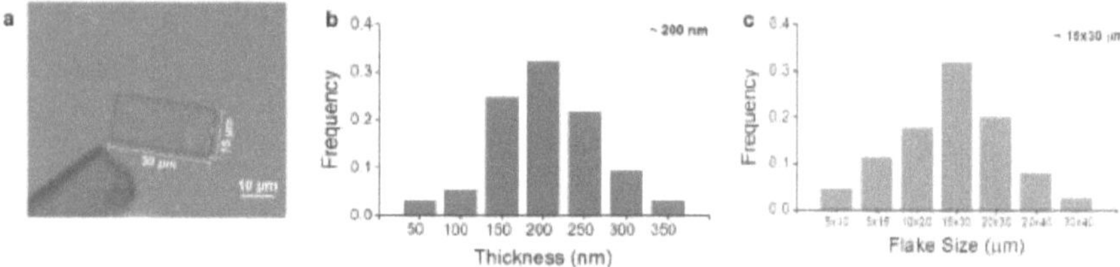

Figure 2.4 (a) A typical optical image of an M-KNbO₃ crystal. (b) Statistical distribution of the thickness of M-KNbO₃ crystals. (c) Statistical distribution of the flake size of M-KNbO₃ crystals.

Figure 2.5a presents a high-resolution transmission electron microscopy (HRTEM) image that was captured along the [001] direction of the M-KNbO₃ crystal after an accurate calibration of sample motion. A high-magnification HRTEM image (Figure 2.5b) and its contrast transfer function (CTF)-calibrated image (Figure 2.5c) were compared to calculate the defocus value of the objective lens (-70 nm), which was used to calibrate the HRTEM image in Figure 2.5a. The

HRTEM image of the crystal after defocus calibration (Figure 2.5d) shows the atomic arrangement

of an M-KNbO$_3$ crystal, which is indicative of the orthorhombic crystal structure of the M-KNbO$_3$

crystals.

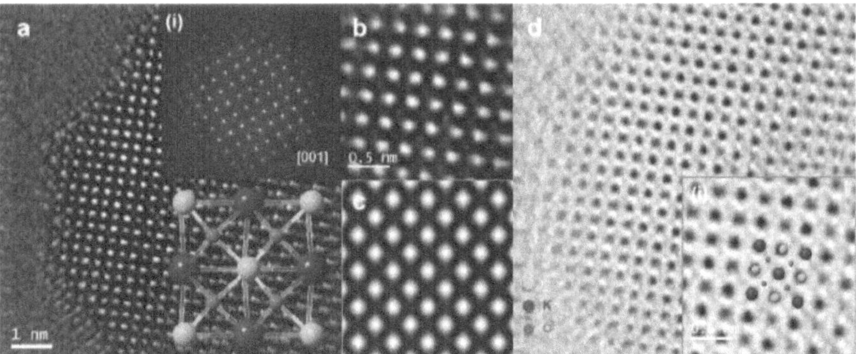

Figure 2.5 (a) High-resolution TEM image of M-KNO crystals. Inset (i): Selected area electron diffraction pattern taken along the [001] direction of a crystal, which signifies the formation of the orthorhombic phase. (b) A high-magnification HRTEM image of M-KNbO$_3$ crystal. (c) Simulated image for calculating the defocus value based on b). (d) The contrast transfer function (CTF)-corrected HRTEM image of M-KNbO$_3$. Inset (i): Atomic arrangement of K, Nb, and O.

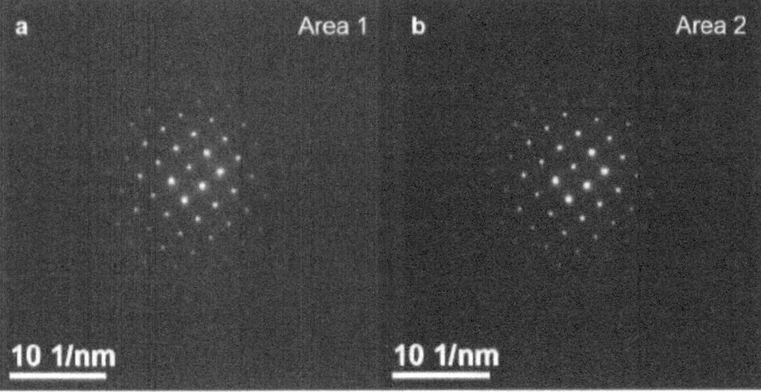

Figure 2.6 Selected area electron diffraction (SAED) patterns of an M-KNbO$_3$ crystal captured from a) area 1 and b) area 2.

Notably, several areas on the same crystal were selected to perform electron diffraction, and all patterns contain one single orthorhombic reciprocal lattice, which evidences the formation of a single $KNbO_3$ crystal (Figure 2.6).

2.2 Process Investigation

To gain insight into the phase evolution process during the reaction, intermediate products obtained after different reaction times were collected and analyzed by XRD. Figure 2.7a-c depict the XRD patterns of products obtained after hydrothermal reaction times ranging from 0 to 48 hours.

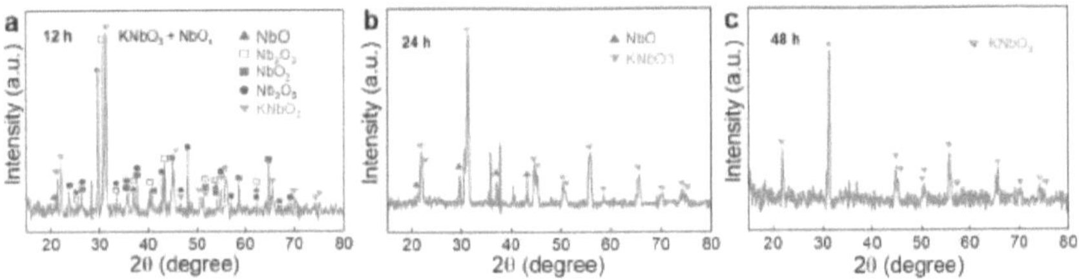

Figure 2.7 X-ray diffraction patterns obtained after different hydrothermal reaction times: (a) 12 hours, (b) 24 hours, and (c) 48 hours.

After the hydrothermal treatment in the KOH aqueous solution, the Nb_2C MXene could be completely converted into $KNbO_3$ crystals. As Figure 2.7a displays, after 12 hours of hydrothermal reaction time, the hydroxyl groups first oxidized the Nb_2C MXene to niobium oxides that include NbO, Nb_2O_3, NbO_2, and Nb_2O_5. Subsequently, in the alkalization process, K^+ reacted with the oxides, which initiated the formation of the $KNbO_3$ phase[29-30]. Figure 2.7b shows that the product that was obtained after 24 hours of reaction time consists of NbO and $KNbO_3$, which indicates that the oxidation and alkalization reactions favor the formation of $KNbO_3$. Figure 2.7c illustrates that only pure $KNbO_3$ was observed after 48 hours of hydrothermal reaction.

For comparison, we synthesized $KNbO_3$ crystals using commercial, non-layered niobium carbide (NbC) powder (Figure 2.8a) and KOH as precursors, and SDS as surfactant. The SEM images reveal cube-like morphology of non-uniform size for the $KNbO_3$ products (Figure 2.8b), which lack the highly textured morphology of MXene-derived $KNbO_3$. In addition, we have investigated the morphologies of intermediate products obtained during the hydrothermal synthesis of M-$KNbO_3$. Figure 2.8c shows an optical image of a single monoclinic Nb_2O_5 crystal (P 2/m) and hexagonal NbO_2 ($P6_3$/mmc) crystal, which both inherit the two-dimensional character of the Nb_2C MXene flake. Thus, we believe that both oxidation and alkalization are initiated on the Nb_2C MXene substrates. In order to investigate the effect of SDS on the morphology of M-$KNbO_3$ crystals, we synthesized crystals under the same conditions without adding SDS surfactant. The final products turn out to be irregular crystals as shown in Figure 2.8d in Supporting Information. Furthermore, only needle-like or rod-like $KNbO_3$ crystals have been hydrothermally synthesized by using Nb_2O_5, KOH as precursors[31], even when adding SDS as surfactant[16]. These results indicate that the successful synthesis of symmetrically-shaped, large-sized, and c-axis oriented $KNbO_3$ crystals is possible tha2nks to the two-dimensional character of MXene flakes and shape-determining effect of SDS. For the first time, we produce such large $KNbO_3$ crystals by hydrothermal synthesis with uniform shape, which makes them useful for practical electro-optic applications.

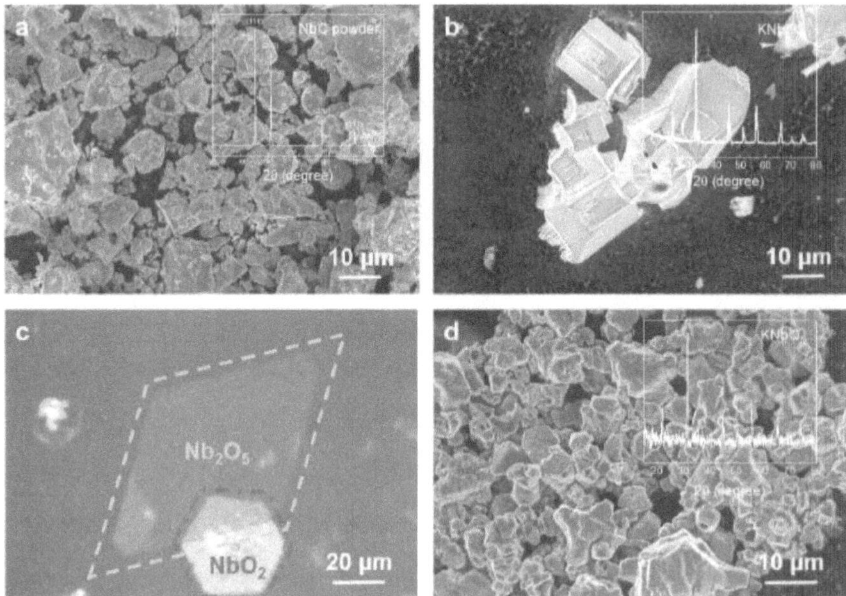

Figure 2.8 a) SEM image of commercial niobium carbide powder. Inset: XRD of commercial NbC powder. b) A typical SEM image of cube-like products by using commercial niobium carbide powder and KOH as precursors, and SDS as surfactant. Inset: XRD of cube-like KNbO₃ crystals. c) An optical image of a single monoclinic Nb_2O_5 crystal (P 2/m) and hexagonal NbO_2 (P6₃/mmc) crystal by hydrothermal synthesis using MXene and KOH as precursors and SDS as surfactant at 190 °C for 12 h. d) A SEM image of product by hydrothermal synthesis using MXene and KOH as precursors without adding SDS as surfactant at 190 °C for 48 h. Inset: XRD of irregular KNbO₃ crystals.

2.3 Ferroelectric Measurements

2.3.1 Device Fabrication

The results of this study confirm that the crystal structure of M-KNbO₃ is orthorhombic, so M-KNbO₃ should be ferroelectric at room temperature. To perform ferroelectric measurements, Pt/M-KNbO₃/Pt two-terminal devices were fabricated by a focused ion beam (FIB) microscope.

2.3.2 Remnant Polarization Measurements

Figure 2.9a offers the current density *versus* electric field (J-E) curve of a Pt/M-KNbO$_3$/Pt device, which reflects a switching current density when electric field is swept at ±50 kV/cm. Figure 2.9b provides the capacitance *versus* electric field (C-E) curve of the Pt/M-KNbO$_3$/Pt device, which presents typical butterfly-type hysteresis over the same electric field range at the frequency of 10 Hz. Both J-E and C-E curves indicate switchable polarization in the M-KNbO$_3$ crystals. Furthermore, ferroelectric hysteresis loops were measured from the same device at 10 Hz as shown in Figure 2.9c[32-33]. Evidently, M-KNbO$_3$ exhibits well-saturated ferroelectric hysteresis loops with a saturation polarization (P$_s$) of ~ 21 µC/cm^2, a remnant polarization (P$_r$) of ~ 17 µC/cm^2, and a coercive field (E$_c$) of ~50 kV/cm. The remnant polarization of M-KNbO$_3$ crystal is close to the best result reported for ceramic KNbO$_3$ (~18 µC/cm^2)[14] (Table 2.1).

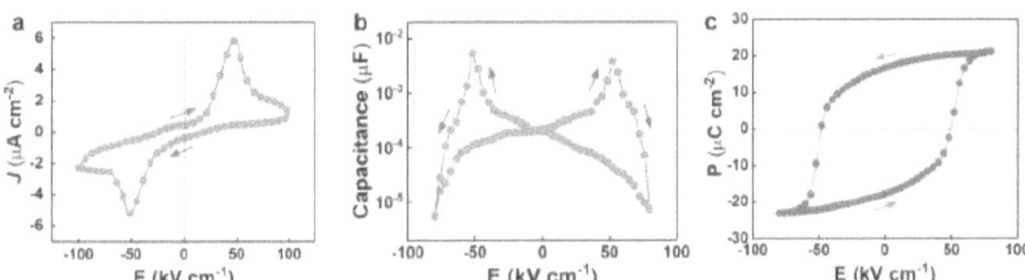

Figure 2.9 Ferroelectric measurement of M-KNbO$_3$ crystals. a) The J-E loop of a Pt/M- KNbO$_3$/Pt device measured with an electric field ranging from -50 kV/cm to 50 kV/cm. b) The C-E loops of Pt/M- KNbO$_3$/Pt device measured with an electric field ranging from -50 kV/cm to 50 kV/cm and an ac voltage of 200 mV. c) The ferroelectric hysteresis loop of an M-KNbO$_3$ crystal measured over the same electric field.

Table 2.1 Ferroelectric Properties of various $KNbO_3$ films, ceramic, and crystals compared with our MXene derived $KNbO_3$.

Forms	Methods	P_r ($\mu C/cm^2$)	E_c (kV/cm)	References
$KNbO_3$ ceramics	Sintering	18	18.3	*J. Am. Ceram. Soc.* **2005**, *88*, 1754.
$KNbO_3$ ceramics	Sintering	7	11	*J. Am. Ceram. Soc.* **1967**, *50*, 329.
$KNbO_3$ films	Hydrothermal	5	100	*Jpn. J. Appl. Phys.* **2014**, *53*, 09PA10.
$KNbO_3$ plates	Hydrothermal	17	50	Present work

2.3.3 Piezoresponse Measurements

Piezoresponse force microscopy measurements were performed to further verify the existence of ferroelectricity in M-$KNbO_3$ crystals (Figure 2.10). We conducted local switching tests by applying a bias between the conductive PFM tip and the platinum-coated silicon substrate. The well-defined butterfly loops of the PFM amplitude signals and the distinct 180-degree switching of the phase signals further corroborate the presence of ferroelectricity in M-$KNbO_3$ crystals. Compared to randomly oriented polycrystalline materials, layer structured ferroelectrics are preferred for ferroelectric and piezoelectric applications due to their lower anisotropy and larger number of permissible orientations for the spontaneous polarization[34-35]. Templated grain growth is one of the promising routes for fabricating textured ferroelectric ceramics with improved functional properties[36]. The platelet-like MXene-derived $KNbO_3$ crystals reported here can be used to fabricate textured $KNbO_3$ ceramics with advanced ferroelectric, piezoelectric, and dielectric properties due to their high aspect ratio and robust ferroelectricity.

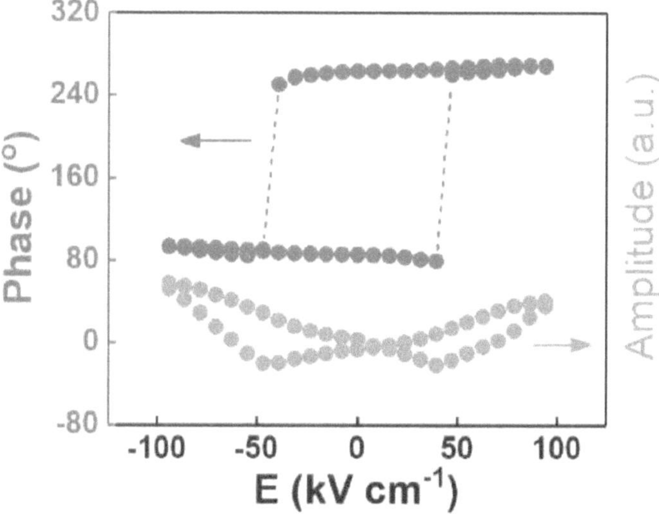

Figure 2.10 The corresponding PFM phase (dark cyan) and amplitude (orange) hysteresis loops during the switching process.

2.4 Conclusion

To summarize, we have demonstrated that two-dimensional MXenes can be utilized as precursors to synthesize ferroelectric single crystals for the first time. As an example, we synthesized orthorhombic $KNbO_3$ ferroelectrics from Nb_2C MXene. Materials and electrical characterization confirm the crystal structure and ferroelectricity of $KNbO_3$. The MXene-derived $KNbO_3$ crystals exhibit a saturation polarization, a remnant polarization, and a coercive field of ~ 21 $\mu C/cm^2$, ~ 17 $\mu C/cm^2$, and ~50 kV/cm, respectively. Our discovery illustrates that the two-dimensional character of MXene can be exploited to grow ferroelectric crystals with specific orientations.

REFERENCES

[1]   Q. Peng, J. Guo, Q. Zhang, J. Xiang, B. Liu, A. Zhou, R. Liu, Y. Tian, *J. Am. Chem. Soc.* **2014**, *136*, 4113.

[2]   S. Tu, Q. Jiang, X. Zhang, H. N. Alshareef, *ACS Nano* **2018**, *12*, 3369.

[3]   X. Yu, Y. Li, J. Cheng, Z. Liu, Q. Li, W. Li, X. Yang, B. Xiao, *ACS Appl. Mater. Interfaces* **2015**, *7*, 13707.

[4]   H. Liu, C. Duan, C. Yang, W. Shen, F. Wang, Z. Zhu, *Sensors Actuators, B Chem.* **2015**, *218*, 60.

[5]   J. Xuan, Z. Wang, Y. Chen, D. Liang, L. Cheng, X. Yang, Z. Liu, R. Ma, T. Sasaki, F. Geng, *Angew. Chemie - Int. Ed.* **2016**, *55*, 14569.

[6]   I. Grinberg, D. V. West, M. Torres, G. Gou, D. M. Stein, L. Wu, G. Chen, E. M. Gallo, A. R. Akbashev, P. K. Davies, J. E. Spanier, A. M. Rappe, *Nature* **2013**, *503*, 509.

[7]   J. C. Baumert, C. Walther, P. Buchmann, H. Kaufmann, H. Melchior, P. Günter, *Appl. Phys. Lett.* **1985**, *46*, 1018.

[8]   M. Zgonik, R. Schlesser, I. Biaggio, E. Voit, J. Tscherry, P. Günter, *J. Appl. Phys.* **1993**, *74*, 1287.

[9]   T. N. & M. N. Yasuyoshi Saito, Hisaaki Takao, Toshihiko Tani, Tatsuhiko Nonoyama, Kazumasa Takatori, Takahiko Homma, *Phys. Status Solidi B* **2004**, *432*, 1.

[10]   Y. Nakayama, P. J. Pauzauskie, A. Radenovic, R. M. Onorato, R. J. Saykally, J. Liphardt, P. Yang, *Nature* **2007**, *447*, 1098.

[11]   S. De Zhong, *Prog. Cryst. Growth Charact. Mater.* **1990**, *20*, 161.

[12]   B. T. Matthias, J. P. Remeika, *Phys. Rev.* **1951**, *82*, 727.

[13]   G. K. L. Goh, F. F. Lange, C. G. Levi, *J. Mater. Res.* **2003**, *18*, 338.

[14]   H. Birol, D. Damjanovic, N. Setter, *J. Am. Ceram. Soc.* **2005**, *88*, 1754.

[15]   A. Nazeri-eshghi, A. X. Kuang, J. D. Mackenzie. *J. Mater. Sci.* **1990**, *25*, 3333.

[16]   G. Z. Wang, S. M. Selbach, Y. D. Yu, X. T. Zhang, T. Grande, M. A. Einarsrud, *Cryst. Eng. Comm.* **2009**, *11*, 1958.

[17]   H. Hayashi, Y. Hakuta, Y. Kurata, *J. Mater. Chem.* **2004**, *14*, 2046.

[18]   B. Li, Y. Hakuta, H. Hayashi, *J. Supercrit. Fluids* **2005**, *35*, 254.

[19]   G. H. HAERTLING, *J. Am. Ceram. Soc.* **1967**, *50*, 329.

[20]   K. J. Kormondy, Y. Popoff, M. Sousa, F. Eltes, D. Caimi, M. D. Rossell, M. Fiebig, P. Hoffmann, C. Marchiori, M. Reinke, M. Trassin, A. A. Demkov, J. Fompeyrine, S. Abel, *Nanotechnology* **2017**, *28*, 075706.

[21]   C. Bi, Q. Wang, Y. Shao, Y. Yuan, Z. Xiao, J. Huang, *Nat. Commun.* **2015**, *6*, 1.

[22]   D. Bai, J. Zhang, Z. Jin, H. Bian, K. Wang, H. Wang, L. Liang, Q. Wang, S. F. Liu, *ACS Energy*

Lett. **2018**, *3*, 970.

[23]   A. Byeon, A. M. Glushenkov, B. Anasori, P. Urbankowski, J. Li, B. W. Byles, B. Blake, K. L. Van Aken, S. Kota, E. Pomerantseva, J. W. Lee, Y. Chen, Y. Gogotsi, *J. Power Sources* **2016**, *326*, 686.

[24]   H. Lin, S. Gao, C. Dai, Y. Chen, J. Shi, *J. Am. Chem. Soc.* **2017**, *139*, 16235.

[25]   D. G. Bozinis, *Phys. Rew. B* **1976**, *13*, 3109.

[26]   J. W. Liu, G. Chen, Z. H. Li, Z. G. Zhang, *Int. J. Hydrog. Energy* **2007**, *32*, 2269.

[27]   S. Yang, Y. Hu, S. Wang, H. Gu, Y. Wang, *Adv. Cond. Matter. Phys.* **2013**, *2013*.

[28]   C. Fischer, *Phys. Stat. Sol. A* **1993**, *137*, 247.

[29]   S. Isobe, K. Kudoh, S. Hino, K. Hara, N. Hashimoto, S. Ohnuki, *Appl. Phys. Lett.* **2015**, *107*.

[30]   S. L. Skjærvo, K. Høydalsvik, A. B. Blichfeld, M.-A. Einarsrud, T. Grande, *R. Soc. Open Sci.* **2018**, *5*.

[31]   A. Magrez, E. Vasco, J. W. Seo, C. Dieker, N. Setter, L. Forro, *J. Phys. Chem. B* **2006**, *110*, 58.

[32]   S. V. Kalinin, N. Setter, A. Kholkin, *MRS Bulletin* **2009**, *34*, 634.

[33]   N. Balke, I. Bdikin, S. V. Kalinin, A. Kholkin, *J. Am. Ceram. Soc.* **2009**, *92*, 1629.

[34]   C. Reitz, P. M. Leufke, H. hahn, T. Brezesinski, *Chem. Mater.* **2014**, *26*, 2195.

[35]   K. H. Wong, C. L. Mak, C. L. Choy, Y. H. Zhang, *J. Vac. Sci. Technol. A* **2000**, *18*, 2412.

[36]   H. Amorin, A. L. Kholkin, M. E. V. Costa, *Mat. Res. Bull.* **2008**, *43*, 1412.

Chapter 3- Nb₂CT$_x$ MXene-derived LiNbO₃ Photoluminescent Ferroelectric Crystals

MXenes have recently been used to grow highly textured potassium niobate ferroelectric crystals. Herein, the versatility of MXenes is further demonstrated by growing ferroelectric and luminescent lithium niobate (LiNbO₃) crystals from niobium carbide MXene (Nb₂CT_x). The formation of high-aspect-ratio LiNbO₃ rhombic crystals was confirmed by extensive structural analysis. The ferroelectricity of MXene-derived LiNbO₃ was verified, using standard ferroelectric, dielectric measurements. In addition, for the first time, we synthesized Pr^{3+}-doped LiNbO₃ crystals with simultaneous visible photoluminescence (PL) from Nb₂CT_x MXene. Our work demonstrates that it is possible, by using the two-dimensional (2D) character of MXenes, to fabricate high aspect ratio and well-oriented photoluminescent ferroelectric crystals for advanced optoelectronic applications.

Over the past decades, LiNbO₃ has garnered significant interest from materials scientists because of its remarkable ferroelectricity, high Curie temperatures, excellent photorefractive and photovoltaic properties, and large electro-optic and acousto-optic coefficients. As a human-made material, it has a rather wide range of applications in actuators[1-2], nonvolatile holographic storage[3], acoustic tweezers[4], waveguides[5], and many more. The massive potential of LiNbO₃ has stimulated research on the synthesis of ceramics[6], nanoparticles[7], and films[8]. However, the synthesis of plate-like LiNbO₃ ferroelectric crystals are not yet reported. LiNbO₃ crystals are particularly interesting because they can be integrated with semiconductor electronics and optoelectronics devices. MXene-based well-orientated LiNbO₃:RE^{3+} crystals would find many applications, such as waveguide lasers[9], mechano-luminescence[10-11], and upconversion light emission[12]. In this study, we illustrate the preparation of plate-like textured $LiNbO_3$ crystals, using accordion-like Nb_2CT_x MXene as a precursor. The rhombic crystal structure (space group *R3C*) of the MXene-based $LiNbO_3$ (M-LN) was verified by both X-ray diffraction (XRD) analysis, and selected- area electron diffraction (SAED), in conjunction with high-resolution transmission electron microscopy (HRTEM). The robust ferroelectricity of M-LN was confirmed by the existence of a switching phase and a polarization. Furthermore, well-orientated photoluminescent MXene-based $LiNbO_3:Pr^{3+}$ ($M-LN:Pr^{3+}$) crystals were synthesized, following a unique chemical process.

3.1 Hydrothermal Synthesis

Figure 3.1a illustrates the synthesis process of Nb_2CT_x-based $LiNbO_3$ crystals. Ascribed to the weak Al-related binding energy in the Nb_2AlC MAX phase[12], Al atoms were selectively etched using a specific acid (e.g. HF) to prepare Nb_2CT_x MXene, which was then used to fabricate the final product, i.e. $LiNbO_3$ crystals. Figure 3.1b shows the crystal structures of the reactant Nb_2CT_x MXene, intermediate products niobium oxides (Nb_xO_y), and the final product $LiNbO_3$, respectively. This was achieved by first oxidizing Nb_2CT_x MXene flakes in a LiOH aqueous solution to form various types of niobium oxides, which we subsequently alkalized to a Li^+ ions in the final M-LN crystals.

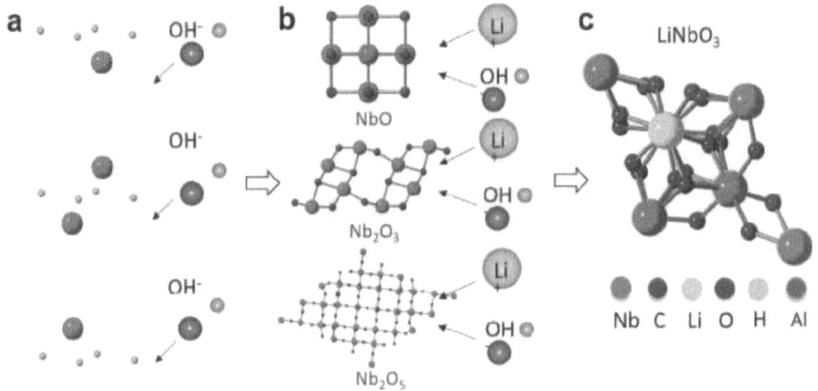

Figure 3.1 Crystal structures of (a) the Nb_2CT_x MXene, (b) intermediate products Nb_xO_y, and (c) the final product M-LN.

Figure 3.2b shows the layered morphology of Nb_2C MXene obtained by etching out the Al layer from the Nb_2AlC MAX phase (Figure 3.2a). Compared with the XRD patterns of the Nb_2AlC MAX phase, the 2θ value of the characteristic (002) peak of Nb_2CT_x MXene shifts from 13° to 5°, indicating an increase of the d-spacing in the Nb_2CT_x MXene phase due to the removal of Al atoms (Figure 3.2c).

Figure 3.2 (a) SEM image of Nb_2CT_x MXene after etching from Nb2AlC MAX phase. (b) XRD pattern of Nb_2AlC MAX phase and Nb_2CT_x MXene.

An analysis of the X-ray diffraction patterns of the products under different lengths of reaction time (Figures 3.3a-b) is performed in order to investigate the process of hydrothermal chemical reaction. For the 24 h-reaction group, the characteristic peaks of NbC, NbO, Nb_2O_5 and $LiNbO_3$ were observed, which clearly conveyed that the Nb_2CT_x MXene was first oxidized to niobium oxides on the way to form $LiNbO_3$ crystals (Figure 3.3a). For the 96 h-reaction group, only characteristic XRD peaks of LiNbO3 crystals were detected in the final product (Figure 3.3b).

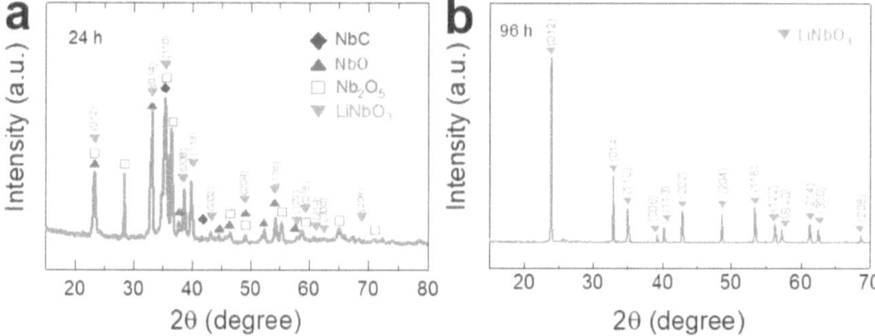

Figure 3.3 XRD patterns of final products after (a) 24 hours, (b) 96 hours chemical reaction.

The existence of niobium oxides was further confirmed by electron microscopy characterization. Figure 3.4a shows a high resolution transmission electron microscopy (HRTEM) image of cubic NbO phase, and the inset (i) confirms its Pm-3m symmetry[13] by selected area electron diffraction (SAED) pattern. In addition, Figure 3.4b depicts the HRTEM image and SAED of a monoclinic Nb_2O_5 crystal (C 2/m)[14].

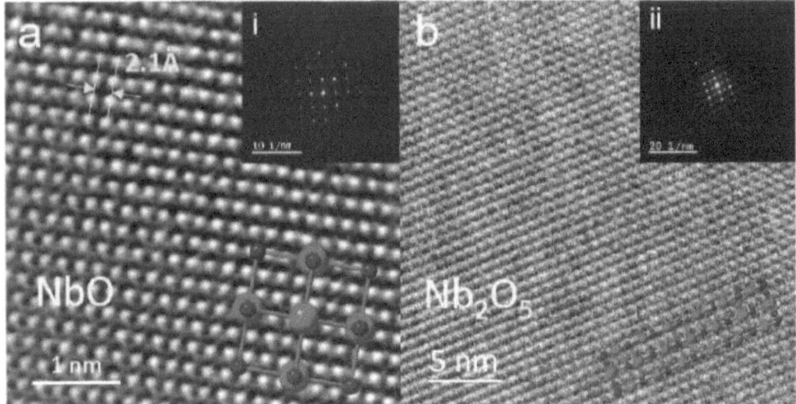

Figure 3.4 (a) HRTEM image of NbO crystal. Inset (i) Selected area electron diffraction of NbO. (b) HRTEM image of Nb_2O_5 crystal. Inset (i) Selected area electron diffraction of Nb_2O_5.

Furthermore, we found out that both the two-dimensional character of Nb_2CT_x MXene and sodium dodecyl sulfate (SDS) surfactant played critical roles in the nucleation and growth of high aspect ratio $LiNbO_3$ crystals during the experimental process. Figure 3.5a-c present the final products obtained by using NLS (commercial NbC powder, LiOH, and SDS), ML (MXene and LiOH), and MLS (MXene, LiOH, and SDS) as reaction combinations, respectively. Figure 3.5a shows that only rhombic-like $LiNbO_3$ crystals were derived from the NLS reaction combination, while irregular products were obtained from the ML group (Figure 3.5b). Only in the case of MLS mixture, the final product presented to be symmetric shaped and large sized $LiNbO_3$ crystal (Figure 3.5c).

Figure 3.5 (a) A typical SEM image of rhombic $LiNbO_3$ crystal by hydrothermal synthesis using commercial NbC powder and LiOH as precursors and SDS as surfactant at 260 °C for 48 h. (b) A SEM image of product by hydrothermal synthesis using MXene and LiOH as precursors without adding SDS as surfactant at 260 °C for 48 h. (c) A typical optical image of plate-like $LiNbO_3$ crystal by using Nb_2C MXene and LiOH as precursors, and SDS as surfactant at 260 °C for 48 h.

Figure 3.6a displays a SEM image of typical Nb_2CT_x-derived $LiNbO_3$ (M-LiNbO₃) crystals obtained using MLS reactants and for a reaction time of 96 hours. Figure 3.7 shows the average size (25 μm * 40 μm) and thickness (150 nm) of M-LN crystals. $LiNbO_3$ crystals are expected to have a R3C symmetry at room temperature, which is verified by XRD analysis[28] (Figure 3.6 b). The microstructure of M-LN crystals was further investigated using electron microscopy. Figure 3.6d shows a high resolution TEM image of a [100] oriented M-LN crystal (c = 13.86 Å), and Figures 3.6e-f respectively represent the experimental and computed selected area electron diffraction pattern along the same [100] direction, which further confirms the rhombic crystal structure of the M-LN crystals[15]. Figure 3.6c displays the Raman spectrum of M-LN crystals, the strong peaks at 149 cm^{-1}, 234 cm^{-1}, and 624 cm^{-1} can be assigned to the E_{TO1} mode, E_{TO2} mode, and E_{TO9} mode, respectively. The additional peaks at 255 cm^{-1}, 271 cm^{-1}, 316 cm^{-1}, 333 cm^{-1}, 367 cm^{-1}, 432 cm^{-1}, and 576 cm^{-1} respectively correspond to the modes of E_{TO3}, $A_{1(TO2)}$, E_{TO4}, $A_{1(TO3)}$, E_{TO5}, E_{TO7}, and E_{TO8} [16].

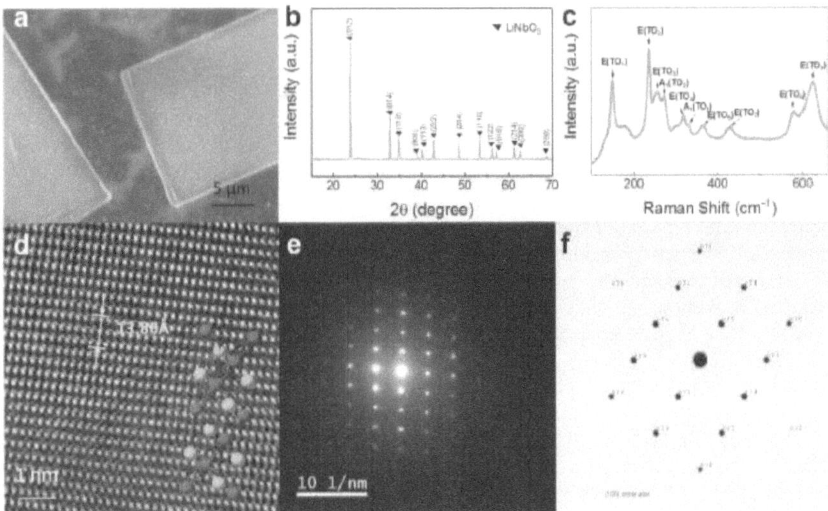

Figure 3.6 a) Scanning electron microscopy image of M-LiNbO3 crystals. b) XRD pattern and c) Raman spectrum of M-LiNbO3 crystals. d) HRTEM image of M-LiNbO3 crystals. e) SAED pattern taken along the [100] direction of M-LiNbO3 crystals. f) Computed SAED pattern taken along the [100] direction of M-LiNbO3 crystals.

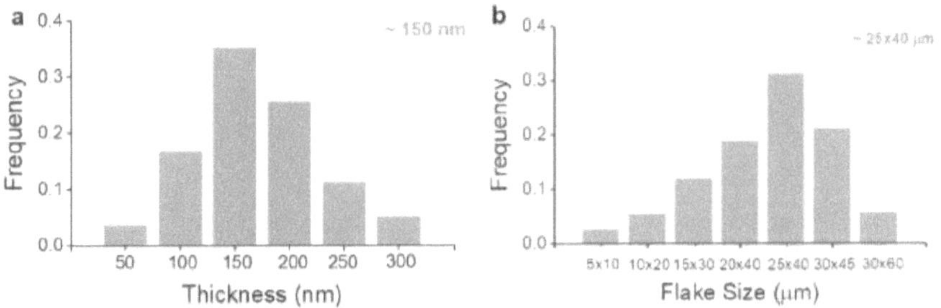

Figure 3.7 (a) Statistical distribution of the thickness of M-LiNbO$_3$ crystals. (b) Statistical distribution of the flake size of M-LiNbO$_3$ crystals.

3.2 Ferroelectricity Measurements

Figure 3.8a presents the ultraviolet-visible absorption spectrum of M-LN crystals, which presents an only absorption at 340 nm within the 200 – 900 nm range. The bandgap of M-LN is evaluated and its measured value is ~ 3.66 eV, which is almost as same as that of ceramic $LiNbO_3$ (~ 3.7 eV)[17]. Furthermore, a Pt/M-LN/Pt UV light photodetector was fabricated using focus ion beam (FIB) microscope. Figure 3.8b shows the UV light response of Pt/M-LN/Pt device after the light illumination under 5V, after it is switched on or off. The excellent ON-OFF switching performance indicates that M-LN can be a good candidate as photodetector, in spite of a relatively large bandgap.

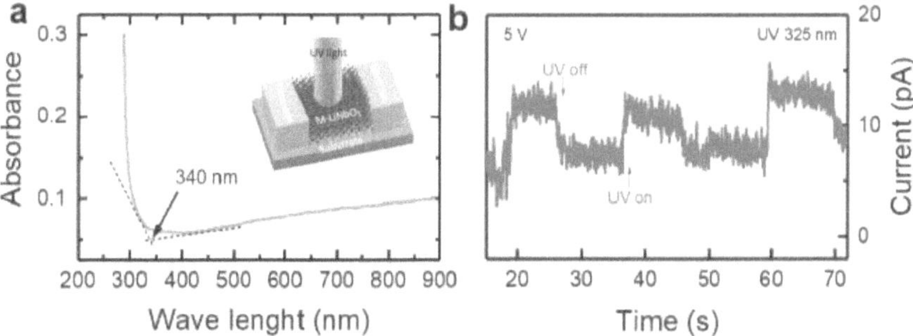

Figure 3.8 (a) The ultraviolet -visible absorbance of $M-LiNbO_3$ crystals. Inset: Schematic illustration of $M-LiNbO_3$ photodetector. (b) The UV light response of $Pt/M-LiNbO_3/Pt$ device after the light illumination is switched on or off measured under 5 V.

In addition, polarization versus voltage (P-V) curve and current density versus voltage (J-V) curve measurements were used to investigate the ferro/piezoelectric properties of R3C symmetric M-LN crystals. The P-V curve of Pt/M-LN/Pt was obtained with voltages ranging from -5 V to 5 V at 10 Hz, which presents a typical ferroelectric hysteresis loop (Figure 3.9). The inserted J-V curve of the same device shows the switching current density at ± 3 V. Both P-V and

J-V curves clearly show the robust ferroelectricity of M-LN crystals; the saturation polarization ($P_s \sim 76~\mu C/cm^2$), remnant polarization ($P_r \sim 68~\mu C/cm^2$), and coercive field (Ec $\sim$ 30 kV/cm) that we obtained were almost as good as the best ones reported in previous studies[18].

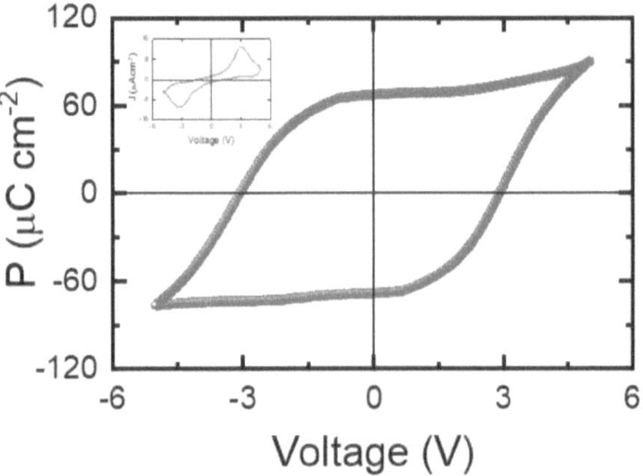

Figure 3.9 The P-V curve of a Pt/M-LiNbO₃/Pt device measured ranging from -5 V to 5 V. Inset: The J-V curve of the same device measured over the same applied external voltage.

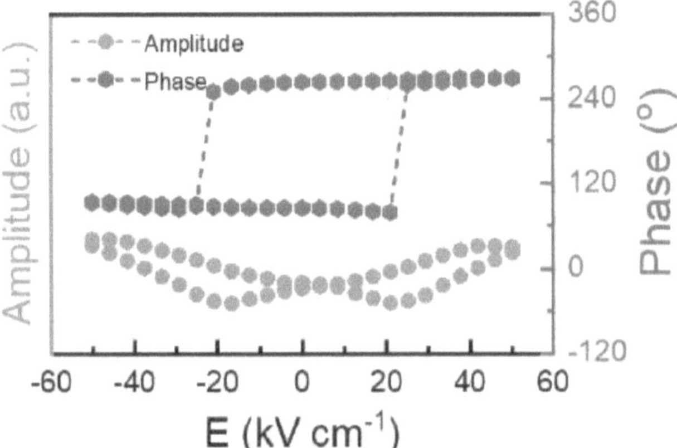

Figure 3.10 The piezoresponse force microscopy of M-LiNbO$_3$ crystal.

In addition, Figure 3.10 shows the phase and amplitude responses under an applied electric field investigated using an atomic force microscope equipped with conductive metal- coated tips. We found that the phase of the [001] oriented M-LN crystal switched from 80° to 260°, at approximately 30 kV/cm, and switched back from 260° to 80° when the applied electric field swept to ~ -30 kV/cm. The 180° phase variation confirmed the presence of ferroelectricity in M-LN crystals, with the butterfly- shaped amplitude loop characteristic of the piezoelectricity of M-LN crystals.

3.3 Photoluminescent Properties

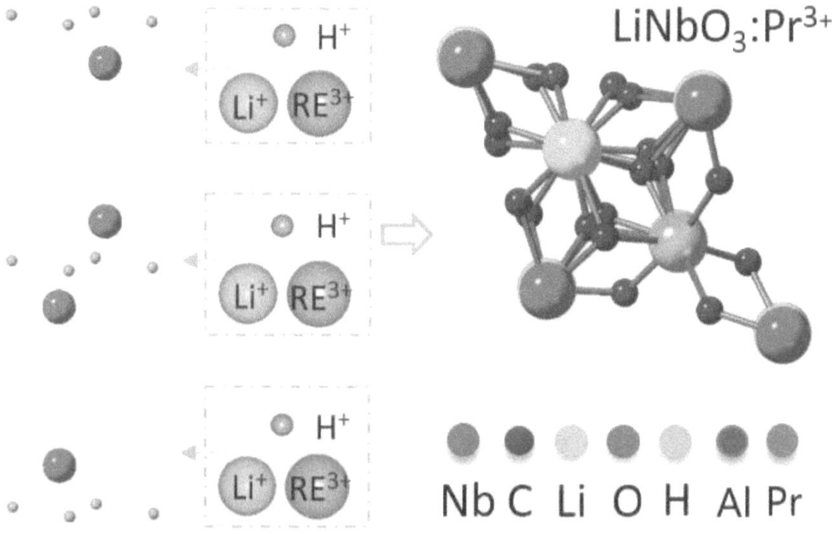

Figure 3.11 Schematic illustrating the synthesis process of Nb$_2$CT_x-based LiNbO$_3$:Pr^{3+} crystals.

The doping of LiNbO$_3$ crystal with rare-earth ions have provided several novel optical properties, such as photoluminescence[19] and mechanoluminescence[10] to the LiNbO$_3$ crystals However, to this day, and to the best of our knowledge, no synthesis of rare-earth-doped LiNbO$_3$ crystals using MXene as precursor using the hydrothermal method was ever reported. Figure 3.11 shows the designed process we used to synthesize LiNbO$_3$:Pr^{3+} crystals with MXene-based trivalent rare-earth ions. Nb$_2$CT_x MXene and Li$_2$CO$_3$ were respectively selected as Nb and Li sources, and Pr^{3+} was selected as a dopant for M-LiNbO$_3$ crystals. Figure 3.12a shows well-oriented Pr^{3+} doped M-LiNbO$_3$ crystals with an average size of ~1 μm^3. In principle, Pr^{3+} ions may substitute for the niobium atoms, lithium atoms, lithium vacancies, and the available free sites [20-21]. Compared to the XRD pattern of undoped LiNbO$_3$ crystals, the slight left shift of the XRD peaks of M-LN:Pr^{3+} crystals (Figure 3.12c) confirms the lattice expansion resulting from the site occupancy by Pr^{3+} dopants. Figure 3.12b displays the red-emitting photoluminescence (PL) of M-LiNbO$_3$:Pr^{3+} crystals under UV light at a wavelength of 325 nm.

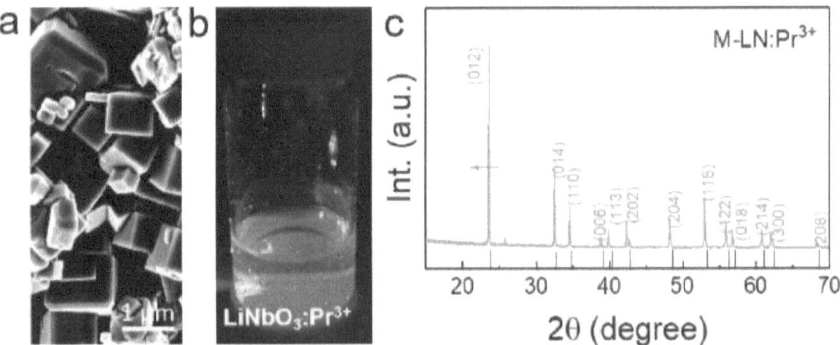

Figure 3.12 (a) SEM image of M-LN:Pr^{3+}, (b) Optical image of M-LN:Pr^{3+} in alcohol under a 325 nm UV light. (c) X-ray diffraction of M-LN:Pr^{3+} crystals.

3.3.1 PL Spectra

Excitation sources with 365 nm, 353 nm, and 274 nm wavelengths were applied to investigate the photoluminescent properties of M-LN:Pr^{3+}. All the PL spectra of M-LN:Pr^{3+} crystals obtained under different excitation conditions (Figure 3.13) consistently exhibited strong peaks at 618 nm and 635 nm.

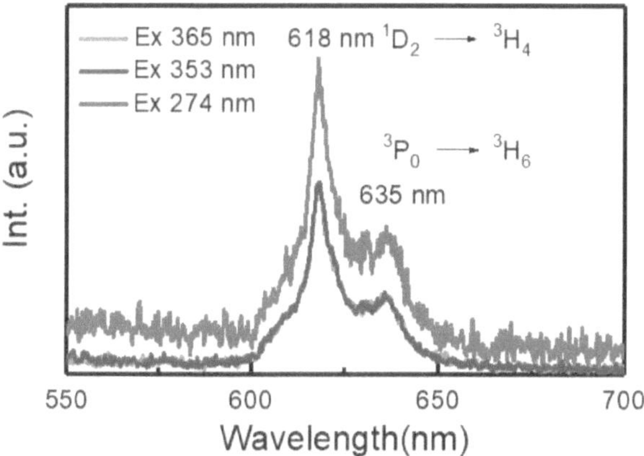

Figure 3.13 Photoluminescence spectra of M-LN:Pr^{3+} crystals at different excitation wavelengths.

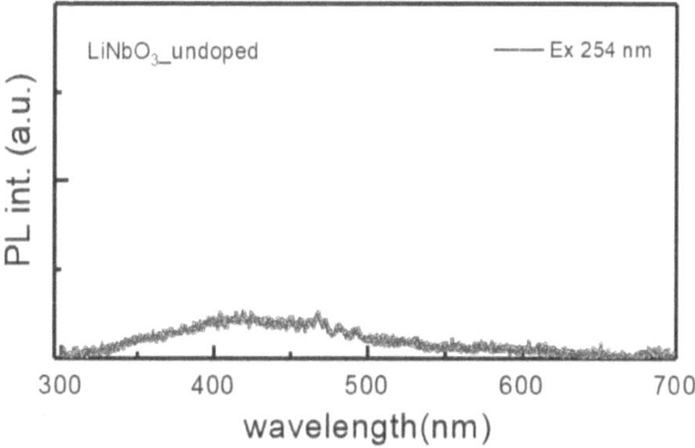

Figure 3.14 Photoluminescence spectrum of undoped $LiNbO_3$ crystals.

For comparison, PL spectra of undoped $LiNbO_3$ crystals were also investigated, indicating no emission peak, for wavelengths ranging from 300 nm to 700 nm (Figure 3.14). By combining the PL spectra with the energy levels diagram of trivalent praseodymium ion (Figure 3.15), we can conclude that the two luminescence peaks (618 nm and 635 nm) of M-LN:Pr^{3+} crystals originate from the dominant $D_2\rightarrow^3H_4$ and $^3P_0\rightarrow^3H_6$ transition of Pr^{3+}, respectively[22].

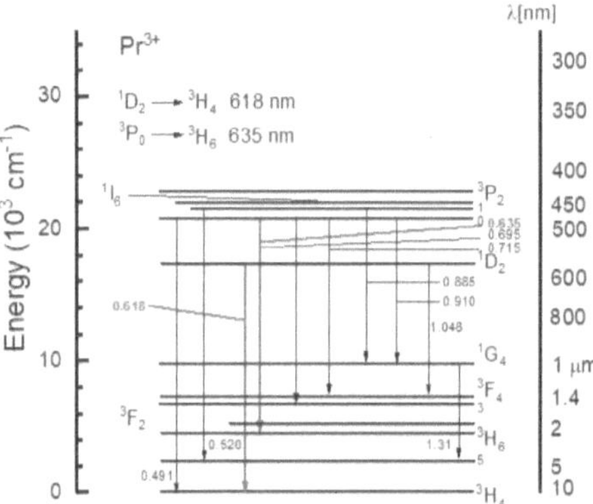

Figure 3.15 Energy level diagrams for the trivalent ions of Praseodymium.

3.3.2 PL Lifetime

Figure 3.16 presents the PL lifetime spectra of M-LN:Pr^{3+} crystals (excitation: 254 nm, emission: 618 nm), which can be well- fitted using a double exponential equation.

$$L = L_0 + a_1 e^{-t/\tau_1} + a_2 e^{-t/\tau_2} \tag{1}$$

where L and L_0 represent the luminescence intensities at times t and 0, a_1 and a_2 are fitting constants, and τ_1 and τ_2 correspond to different lifetime constants, respectively. We also applied the fitting and lifetime constants to calculate the effective lifetime constant (τ_e) of M-LN, as shown in Equation 2:

$$\tau_e = \frac{a_1\tau_1{}^2 + a_2\tau_2{}^2}{a_1\tau_1 + a_2\tau_2} \tag{2}$$

The fitting results showed that lifetime τ_1, τ_2, and τ_e were approximately 1.6 µs, , 20.9 µs, and 1.9 µs, respectively (see Table 3.1 for details).

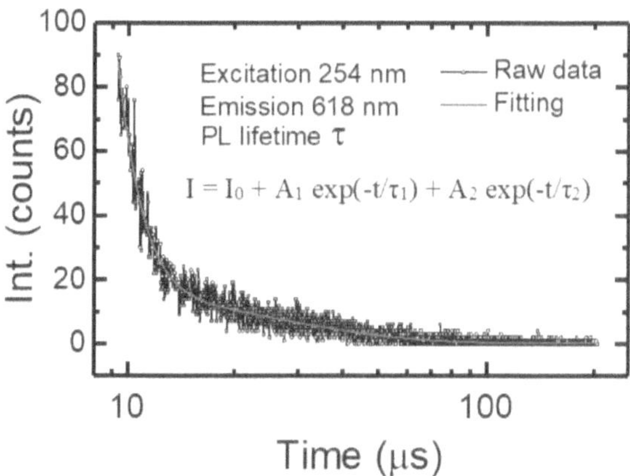

Figure 3.16 PL lifetime spectra of M-LN:Pr^{3+} crystals.

Table 3.1 Fitting parameters of PL lifetime curve.

Model	ExpDec2
Equation	$L = L_0 + a_1\ exp(-t/\tau_1) + a_2\ exp(-t/\tau_2)$
Plot	B
L_0	0.1907
a_1	22041.42272
τ_1	1.62681
a_2	26.75561
τ_2	20.93033
τ_e	1.923649
Reduced Chi-Sqr	2.57399
R-Square(COD)	0.95105
Adj. R-Square	0.951

The double- exponential decay is likely due to the cross-relaxation process of $^3P_0$ + $^3H_4 \rightarrow ^1D_2 + ^3H_6$[24]. We found that the effective PL lifetime of the $^1D_2$ level was approximately 33 μs for 0.04 mol % Pr-doped $LiNbO_3$[25]. The doping concentration of our M-LN:Pr^{3+} crystals was approximately 1.0 mol %, and thus it is reasonable to attribute the shorter luminescent decay of the M-LN:Pr^{3+} crystals due to the well-known concentration quenching effect[24].

3.4 Conclusion

This study presented a methodology to exploit the 2D character of Nb_2CT_x MXenes to fabricate plate-like $LiNbO_3$ crystals with a rhombic crystal structure and a robust ferroelectricity. The saturation polarization, remnant polarization, and the coercive field of M-LN are almost equal to those reported in previous studies. For the first time, we doped trivalent rare-earth ions in the M-LN crystals using a different synthesizing process, and successfully provided the doped $LiNbO_3$ crystals with photoluminescent properties in. Our discovery not only expands the feasibility of deriving oriented ferroelectric crystals using the two-dimensional character of MXenes, but it also offers a new approach for synthesizing MXene-based materials with advanced optoelectronic applications.

REFERENCES

[1]   S. H. Wemple, M. DiDomenico, and I. Camlibel, *Appl. Phys. Lett.* **1968**, *12*, 209.

[2]   S. Kim, V. Gopalan, A. Gruverman, *Appl. Phys. Lett.* **2002**, *80*, 2740.

[3]   L. Hesselink, S. S. Orlov, A. Liu, A. Akella, D. Lande, R. R. Neurgaonkar, *Science* **1998**, *282*, 1089.

[4]   S. Nadupalli, J. Kreisel, T. Granzow, *Sci. Adv.* **2019**, *5*, 9199.

[5]   N. Mkhitaryan, J. Zaraket, N. Kokanyan, E. Kokanyan, M. Aillerie, *Eur. Phys. J. Appl. Phys.* **2019**, *85*, 30502.

[6]   B. D. Wood, V. Mocanu, B. D. Gates, *Adv. Mater.* **2008**, *20*, 4552.

[7]   Y. Wang, X. Y. Zhou, Z. Chen, B. Cai, Z. Z. Ye, C. Y. Gao, J. Y. Huang, *Appl. Phys. A* **2014**, *117*, 2121.

[8]   A. Bartasyte, S. Margueron, T. Baron, S. Oliveri, P. Boulet, *Adv. Mater. Interfaces* **2017**, *4*, 1600998.

[9]   W. Sohler, B. K. Das, D. Dey, S. Reza, H. Suche, R. Ricken, *IEICE TRANS. ELECTRON.* **2005**, *88*, 990.

[10]   A. Feng, P. F. Smet, *Materials* **2018**, *11*, 484.

[11]   D. Tu, C. N. Xu, A. Yoshida, M. Fujihala, J. Hirotsu, X. G. Zheng. *Adv. Mater.* **2017**, *29*, 1606914.

[12]   A. H. Li, Z. R. Zheng, T. Q. Lü, Q. Lü, We. L. Liu, *OPTICS EXPRESS* **2009**, *17*, 3878.

[13]   A. Byeon, A. M. Glushenkov, B. Anasori, P. Urbankowski, J. Li, B. W. Byles, B. Blake, K. L. Van Aken, S. Kota, E. Pomerantseva, J. W. Lee, Y. Chen, Y. Gogotsi, *J. Power Sources* **2016**, *326*, 686.

[14]   A. L. Bowman, T. C. Wallace, J. L. Yarnell, R. G. Wenzel, *Acta Cryst.* **1966**, *21*, 843.

[15]   C. Wang, Q. Su, F. Liu, Y. Qian, G. Zhao, *NanoStruct. Mater.* **1997**, *8*, 163.

[16]   H. D. Megaw, *Acta Cryst.* **1968**, *24*, 583.

[17]   M. Zezulová, M. Jelínek, T. Kocourek, V. Vorlíček, V. Železný1, *Laser Phys.* **2014**, *24*, 025701.

[18]   K. Kam, J. H. Henkel, H. Hwang, *J. Chem. Phys.* **1978**, *69*, 1949.

[19]   S. H. Wemple, M. DiDomenico, and I. Camlibel, *Appl. Phys. Lett.* **1968**, *12*, 209.

[20]   Y. Li, *Phys. Rev. B* **2015**, *91*, 174106.

[21]   A. Lorenzo, H. Jaffrezic, B. Roux, G. Boulon, J. García-Solé, *Appl. Phys. Lett.* **1995**, *67*, 3735.

[22]   A. Lorenzo, L. E. Bausá, J. García Solé, Phys. Rev. B **1995**, *51*, 16643.

[23] F. Chun, B. Zhang, H. Su, H. Osman, W. Deng, W. Deng, H. Zhang, X. Zhao, W. Yang, *J. Lumin.* **2017**, *190*, 69.

[24] W. Ryba-Romanowski, I. Sokólska, S. Gołab, T. Łukasiewicz, *Appl. Phys. Lett.* **1997**, *70*, 686.

Chapter 4- Ti$_3$C$_2$T$_x$ MXene/P(VDF-TrFE-CFE) Composites

MXenes with 2D nature not only were capitalized to synthesize plate-like ferroelectric and photoluminescent a functional materials, but also can be embedded in polymer matrix to fabricate hybrid dielectric nanocomposites, which leads to advanced dielectric enhancement due to the formation of microscopic dipoles at the interface between MXene flakes and polymer matrix. Dielectric polymer nanocomposites are rapidly emerging as novel materials for a number of advanced engineering applications, especially for high energy density capacitor. The energy density of linear dielectrics can be expressed as following:

$$J = \int_0^{E_{\max}} PdE = \int_0^{E_{\max}} \varepsilon_0 \varepsilon_r EdE = \frac{1}{2}\varepsilon_0 \varepsilon_r E^2 \tag{1}$$

J is energy density, ε_r stands for relative dielectric permittivity, and E represents the electric breakdown strength. In order to achieve high energy density, the dielectric candidate should have high dielectric constant and high electric breakdown strength at the same time, as illustrated in Figure 4.1a. P(VDF-TrFE-CFE) has been used for high energy density capacitor contributed to their relatively high dielectric constant/loss ratio, in where various types of reported conductive fillers have been added to form nanocomposites, such as reduced graphene oxide (rGO), carbon nanotube (CNT), and BaTiO$_3$ nanoparticles[1-3] (Figure 4.1b). Compared to other fillers, rGO/P(VDF-TrFE-CFE) nanocomposite shows much higher dielectric permittivity enhancement around percolation threshold and keep a reasonable value of dielectric loss. However, the acceptable window of rGO loading is extremely narrow, which makes it is difficult to fabricate rGO doped nanocomposite with required dielectric properties. Thus, a new type of filler with the performance of low loading sensitivity and high dielectric enhancement becomes quite desirable to optimize dielectric polymer nanocomposites.

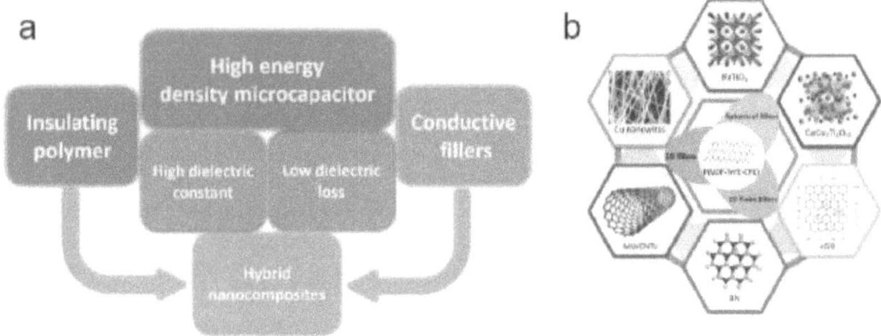

Figure 4.1 a) Materials design that will lead to the optimal performance of high energy density capacitor. b) A summary of composites for high energy density capacitors.

2D $Ti_3C_2T_x$ MXene has large aspect ratio and high electrical conductivity, which make it an excellent filler for polymer composites enhancement. In this chapter, we investigated the dielectric properties of $Ti_3C_2T_x$ MXene/polymer composites. The dielectric permittivity of 2D $Ti_3C_2T_x$ MXene/poly(vinylidene fluoride-trifluoro-ethylene-chlorofluoroehylene) (P[VDF-TrFE-CFE]) reaches up to 10^5 near the percolation limit, which overwhelm all previously reported results of other fillers in the same matrix. When loading MXene in other polymer matrix, as well we observed dielectric enhancement effect. Combined with experimental results and proposed model, we believe that the formation of microscopic dipoles at the surfaces between the MXene sheets and the polymer matrix make a predominant contribution to the dielectric properties enhancement when applying external electric field.

4.1 Characterization of $Ti_3C_2T_x$ MXene

4.1.1 Morphology

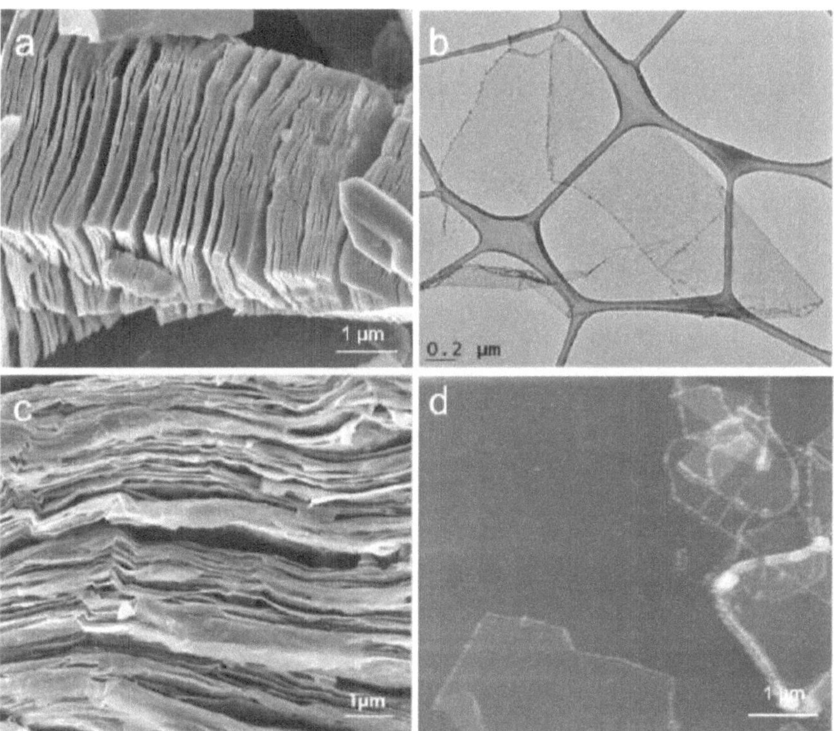

Figure 4.2 Morphology of $Ti_3C_2T_x$ MXene flakes. (a) SEM image of $Ti_3C_2T_x$ MXene after HF acid etching. (a) TEM image of a single $Ti_3C_2T_x$ MXene flake after delamination. (c) Cross-sectional SEM image of $Ti_3C_2T_x$ MXene film after vacuum filtration. (d) Atomic force microscopy image.

Figure 4.2a shows the accordion-like microstructure of $Ti_3C_2T_x$ MXene synthesized from

HF acid etched Ti_3AlC_2 MAX phase. The transmission electron microscopy (TEM) image in

Figure 4.2b depicts the typical morphology of a single-layer $Ti_3C_2T_x$ MXene sheet with the size

of 2*3 μm were obtained after delamination. The foldable edge of MXene flake proves its high flexibility, which is significant to the application of flexible electronic devices. Furthermore, the morphology of $Ti_3C_2T_x$ MXene was further investigated by using atomic force microscopy (AFM). Figure 4.2d shows that the typical size is 3*3 um, and the thickness is ~1.6 nm. In addition, in order to keep the high quality of $Ti_3C_2T_x$ MXene, flesh MXene aqueous suspension was filtrated by suing vacuum to obtain flexible films. Figure 4.2c shows the layered microstructure of filtrated MXene film.

4.1.2 Structure

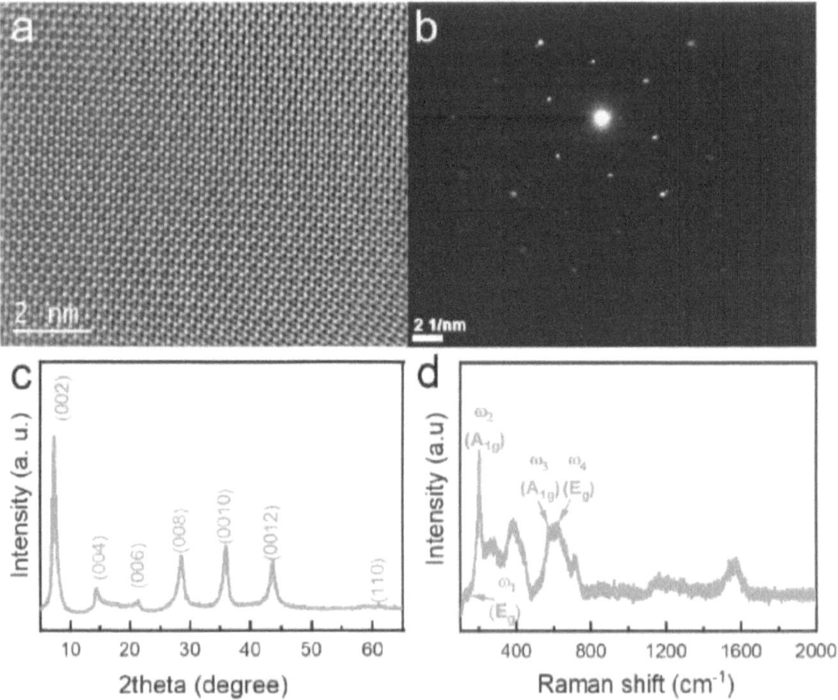

Figure 4.3 Structural characterization of $Ti_3C_2T_x$ MXene. (a) Atomic-resolution TEM image, (b) Selected area electron diffraction (SAED) pattern, (c) XRD pattern, and (d) Raman spectrum.

Figure 4.3a shows a high resolution TEM image of single layer $Ti_3C_2T_x$ MXene, which exhibits hexagonal symmetry of Ti atoms. The hexagonal atomic arrangement of $Ti_3C_2T_x$ MXene was further confirmed by selected area electron diffraction pattern (SAED) (Figure 4.3b). In addition, the MXene sheets were further investigated by using Raman spectroscopy and X-ray diffraction (XRD). The XRD pattern of $Ti_3C_2T_x$ MXene presents a strong (002) characteristic peak at $2\theta = 7.2°$ with FWHM = 0.55°, which indicates a d-spacing of 1.2 nm (Figure 4.3c). Figure 4.3d shows the Raman spectrum of $Ti_3C_2T_x$/CF, which presents two strong peaks at 205 and 721 cm^{-1} contributed to the A_{1g} modes of $Ti_3C_2O_2$. These peaks at 128, 280, 620, and 730 cm^{-1} respectively correspond to the following vibrational modes: E_g of $Ti_3C_2F_2$, E_g of $Ti_3C_2(OH)_2$, E_g of $Ti_3C_2F_2$, and A_{1g} of $Ti_3C_2O_2$. These Raman peaks reveal the existence of functional groups on the $Ti_3C_2T_x$ MXene surface (-OH, -O, and –F)[4].

4.1.3 Surface Chemistry

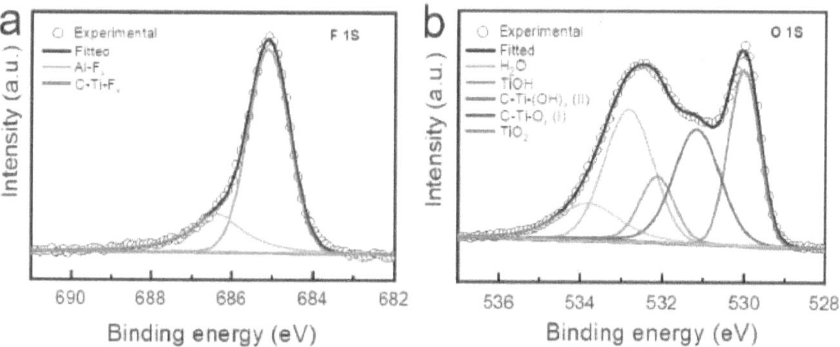

Figure 4.4 Surface chemistry of $Ti_3C_2T_x$ MXene. (a) O 1s and (b) F 1s core levels as obtained by X-ray photoelectron spectroscopy (XPS).

Additionally, the surface chemistry of $Ti_3C_2T_x$ MXene was characterized by using X-ray photoelectron spectroscopy (XPS). Figure 4.4 shows the high-resolution XPS spectra of $Ti_3C_2T_x$ MXene, which indicates the binding energy of O 1s and F 1s. Figure 4.4a shows that the spectra of F 1s region (685.1 eV) can be fitted using symmetric peaks of F-terminated Ti and AlF_x, respectively. Figure 4.4b illustrates that the O 1s peak can be fitted using four symmetric peaks at 530 eV, 531 eV, 532.2 eV, and 532.8 eV, which respectively correspond to TiO_2, C-Ti-O_x, C-Ti-$(OH)_x$, and TiOH[5]. The peak at 533.8 eV can be attributed to H_2O[6].

4.2 Dielectric Enhancement

4.2.1 Sample Preparation

MXene sheets were weighed according to the desired filler loadings, suspended in 3 ml dimethylformamide (DMF), and dispersed *via* shaking for one hour. Subsequently, 300 mg of P(VDF-TrFE-CFE) (61.5/30.2/8.3 mol% Piezotech S.A., France), P(VDF-TrFE-CTFE) (66.5/24.9/8.6 mol% Piezotech S.A., France), P(VDF-TrFE) (70/30 mol % Piezotech S.A., France), and PVP (Sigma-Aldrich, USA) were dissolved into the suspensions under continuous stirring for 30 minutes at 80 °C. The MXene-polymer solutions were then ultrasonicated for two hours, and the dispersion was cast directly onto platinum-coated silicon substrates. The samples were then dried overnight in an oxygen-free glove box and were finally annealed at 70 °C in a vacuum oven for two days. For dielectric characterization, top Ti (100 nm)/Au (50 nm) electrodes were deposited on the films (with an average thickness of 100 um) by electron-beam evaporation through a shadow mask. Figure 4.5 shows the metal-insulator-metal (MIM) capacitor device fabricated using MXene/P(VDF-TrFE-CFE) composite as the dielectric layer.

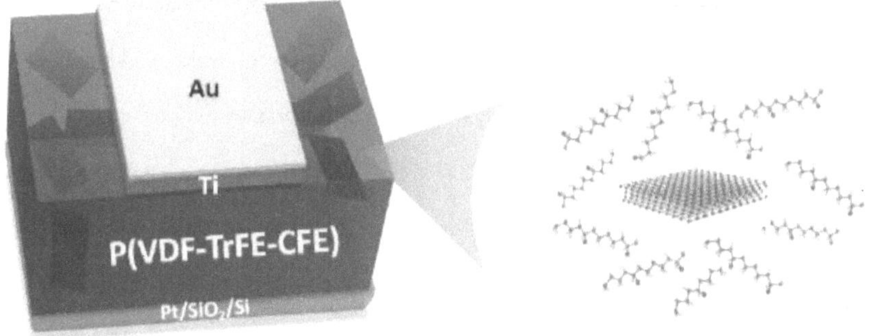

Figure 4.5 Schematic illustration of the capacitor device fabricated using a MXene/P(VDF-TrFE-CFE) composite as the dielectric layer.

The MXene loading concentration of MXene/polymer composites were investigated by using thermogravimetric analysis (TGA). The volume fraction of MXene sheets in P(VDF-TrFE-CFE) matrix can be calculated based on their weight percent, as shown in Table 4.1.

Table 4.1 The weight percent and volume fraction of MXene sheets embedded in P(VDF-TrFE-CFE) matrix.

Wt % MXene	0	1	1.8	4	8	10.7	14.1	15.3	19.5	22.3
Vol % MXene	0	0.45	0.81	1.8	3.6	4.8	6.3	6.9	8.8	10

4.2.2 Dielectric Properties

Figure 4.6a reveals that the dielectric loss also increases with MXene loading. An interesting observation is that the dielectric loss of the MXene/P(VDF-TrFE-CFE) composite increases approximately fivefold (from 0.06 to 0.35) up to 10 wt % MXene loading, while the dielectric constant increases by 25 times over the same composition range. This excellent performance of the composites is best understood through a comparison of the performance of the MXene/P(VDF-TrFE-CFE) composite with published works on dielectric constant enhancement in the same polymers using a variety of fillers. Specifically, Figure 4.6b indicates the dielectric constant and dielectric loss of the P(VDF-TrFE-CFE) polymer in which various types of conductive fillers have been added[7]. The fillers include hydrothermally reduced graphene oxide (HT-rGO)[1], hydrazine-reduced graphene oxide (HZ-rGO)[1], carbon nanotubes (CNTs)[8], copper phthalocyanine (CuPc)[9], polyaniline (PANI)[10], and functionalized graphene nanosheets (FNGS)[11]. The data of the MXene/P(VDF-TrFE-CFE) clearly stand out, as they show the best dielectric constant/loss factor trade-off among all fillers. In fact, at the same dielectric loss as HZ-rGO, (~2), the MXene-based composite shows reasonably higher dielectric constant (15,900 *versus* 12,600).

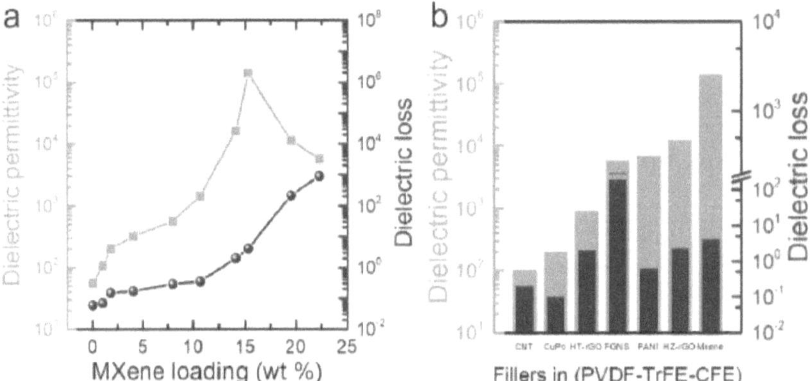

Figure 4.6 (a) Dependence of the permittivity and dielectric loss of the MXene/P(VDF-TrFE-CFE) on MXene content wt % measured at room temperature and 1kHz; (b) Bar charts comparing the maximum dielectric permittivity and corresponding dielectric loss reported in the literature using P(VDF-TrFE-CFE) as a matrix with different conductive fillers.

In addition, at the same loss factor of CuPc composites (0.35), MXene composites achieve a significantly higher dielectric constant (MXene 1,425 *versus* CuPc 75) at 1 kHz. However, Figure 4.6a also illustrates that the permittivity starts to diminish when the MXene loading reaches approximately 15.3 wt%, which is near the percolation limit. The drop in the dielectric constant at higher MXene concentrations is likely due to rising leakage currents caused by increased connectivity between the MXene sheets, which leads to a transition from non-Ohmic to Ohmic conduction[12].

Table 4.2 presents the XRD analysis of MXene/P(VDF-TrFE-CFE) composites with different polymer loadings. Evidently, as the polymer fraction increases, the MXene interlayer spacing expands. One possible explanation for this result is the intercalation of some polymer chains between MXene layers. This intercalation may have been assisted by the functional groups on the MXene surface (*e.g.* F, O, and/or OH), which may interact with atoms (*e.g.* H) on the P(VDF-TrFE-CFE) chains. In fact, polymer chains can reportedly be intercalated between MXene sheets during the in-situ polymerization process of polypyrrole (PPy)[13].

Table 4.2 Diffraction peak angle and interlayer spacing of MXene in P(VDF-TrFE-CFE) matrix with different polymer loading.

Polymer loading (wt %)	2θ	d
96.7	6.45	13.69
93.2	6.49	13.68
89.2	6.52	13.54
86.6	6.56	13.46
0	7.0	12.1

Table 4.3 specifies the dielectric constant, dielectric loss, and conductivity of MXene/P(VDF-TrFE-CFE) composites with different MXene loadings.

Table 4.3 Dielectric permittivity and dielectric loss of MXene/P(VDF-TrFE-CFE) with different MXene loadings.

MXene loading (wt %)	Conductivity (S/m)	Dielectric loss	Dielectric permittivity
0	5.0E-7	0.06	55
1	1.6E-6	0.07	106
1.8	1.08E-5	0.15	202
4.0	9.5E-5	0.17	317
8.0	1.2E-5	0.29	560
10.7	7.0E-4	0.35	1425
14.0	0.015	2.1	15900
15.3	0.14	4.1	139830
19.5	37.4	213	11332
22.3	273	890	5679

Additionally, for comparison, Table 4.4 indicates the maximum dielectric permittivity and corresponding dielectric loss reported in the literature using P(VDF-TrFE-CFE) as the matrix with different conductive fillers.

Table 4.4 The maximum dielectric permittivity and corresponding dielectric loss reported in the literature using P(VDF-TrFE-CFE) as matrix with different conductive fillers.

Conductive filler	Dielectric loss	Dielectric permittivity
CNT	0.2	100
CuPc	0.1	200
HT-rGO	2	900
FGNS	200	5800
PANI	0.6	7000
HZ-rGO	2.2	12500
MXene (14.0 wt %)	2.1	15900
MXene (15.3 wt %)	4	139830

4.2.3 Mechanism of Dielectric Enhancement

The dielectric polarization P is defined to be the bound charges on the capacitor plates per unit area (σ_p), while the total electrical displacement D is defined as the total charge on the capacitor plates (σ_T). For a linear dielectric in which the polarization is linearly proportional to the dielectric constant, as shown in Equation 1-2:

$$E = \frac{\sigma_{free}}{\varepsilon_0} = \frac{\sigma_T - \sigma_P}{\varepsilon_0} = \frac{D - P}{\varepsilon_0} \tag{1}$$

$$D = \varepsilon_0 E + P = \varepsilon_0 E + \chi_e\, \varepsilon_0 E = (1 + \chi_e)\varepsilon_0 E = \varepsilon_0 \varepsilon_r E \tag{2}$$

Figure 4.7 shows that there are four types of polarization in polymer composites, electronic polarization P_e, ionic polarization P_i, dipolar polarization P_{dip}, and interfacial polarization P_{int}, corresponding to their frequency dependences.

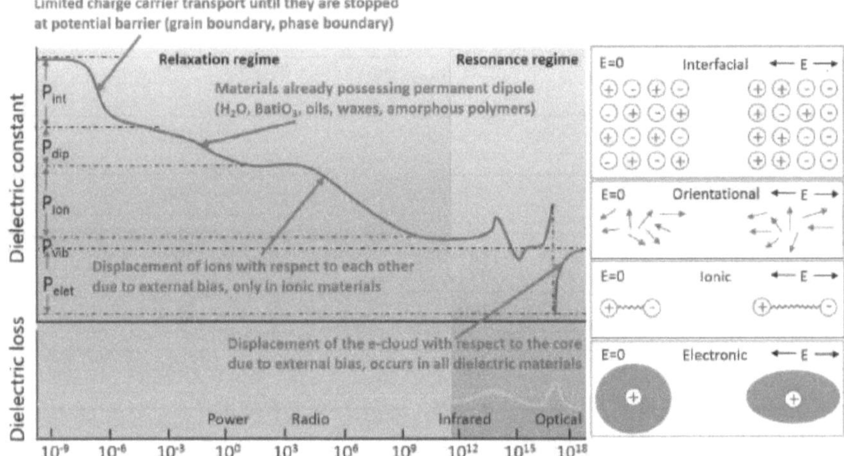

Figure 4.7 Different types of polarizations and their frequency dependences. Here, P_e, P_i, P_{dip}, and P_{int} refer to electronic, ionic, dipolar, and interfacial polarizations. Dielectric constant and corresponding dielectric loss depicted with blue and pink lines, respectively.

In order to study the polarization in MXene/polymer composites, we performed bipolar and unipolar current density vs electric field (J-E) loop measurements on both pristine P(VDF-TrFE-CFE) and MXene/P(VDF-TrFE-CFE) composites[14]. Before either the bipolar or the unipolar loop measurement, the samples were poled negatively by applying an electric field of 2.5kV cm^{-1} for five minutes at room temperature. Figure 4.7a illustrates the first and third *J–E* curves of the pristine P(VDF-TrFE-CFE) sample measured by applying an electric field from -2.5 kV/cm to 2.5 kV/cm, which reveals almost no difference between these two curves. This result demonstrates that the pristine terpolymer P(VDF-TrFE-CFE) behaves as a linear dielectric, which is expected given the paraelectric nature of this polymer. Figure 4.7b displays a sequence of positive unipolar J-E curves (measured on negatively pre-poled terpolymer). As can be seen in Figure 4.7b, inset (i), the current densities of the P(VDF-TrFE-CFE) sample exhibit no obvious change, even after 19 unipolar cycles. This result is not surprising since the terpolymer is not ferroelectric after 70 °C annealing[15], so no switching current can be measured. Figure 4.7b, inset (ii) shows the time-dependent current density of the pristine terpolymer, which was measured by applying a 1.25 kV cm^{-1} electric field and holding it for one hour. Evidently, the current density quickly decreases before stabilizing at a constant value (9.0×10^{-4} μA/cm^2). The results in Figure 4.7a and b clearly demonstrate that pristine P(VDF-TrFE-CFE) exhibits a typical paraelectric behavior and no switching current. The XRD analysis ($2\theta = 16°$ to $22°$) in Figure 4.8 presents only one sharp peak at $2\theta = 18.4°$, which indicates that the paraelectric polymer phase is prominent in both the pristine P(VDF-TrFE-CFE) and the MXene/P(VDF-TrFE-CFE) composite[15].

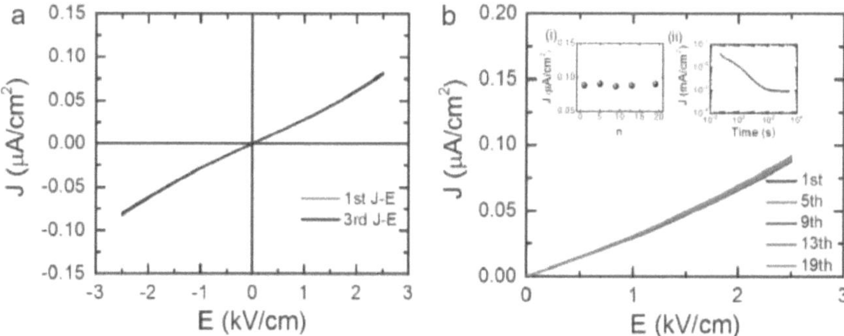

Figure 4.8 Pristine Polymer: (a) The first *J–E* loop and third J-E curve of pristine P(VDF-TrFE-CFE) measured with an electric field of -2.5 kV cm^{-1} to 2.5 kV cm^{-1}; (b) The unipolar *J–E* loops measured with a positive field on a negatively pre-poled pristine P(VDF-TrFE-CFE) sample with an electric field of 2.5 kV cm^{-1}; Inset (i): Variation of the current density at maximum field with the number of cycles (*n*) of the field; Inset (ii): Time-dependent leakage current density measured under a constant field of 1.25 kV cm^{-1} for one hour.

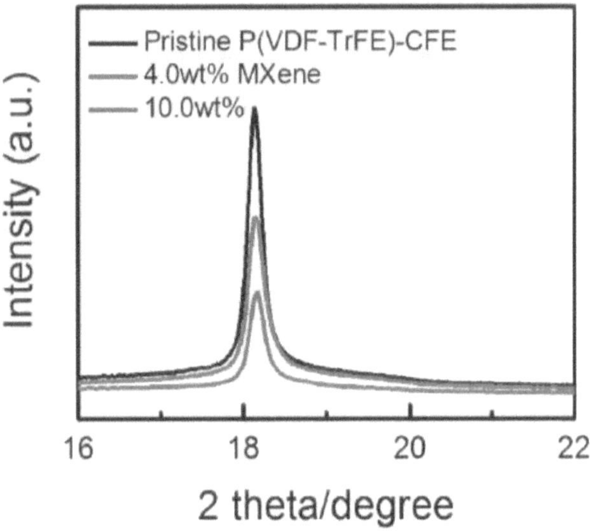

Figure 4.9 XRD pattern of MXene/P(VDF-TrFE-CFE).

Figure 4.9a and b respectively display the bipolar and the unipolar measurement of the MXene/P(VDF-TrFE-CFE) composite. Prior to these measurements, the composite had been poled using negative 2.5kV cm^{-1} for five minutes at room temperature. Figure 4.9a shows the first and third J-E curves measured after negatively poling the sample. These indicate that the introduction of MXene to the terpolymer enables the measurement of a much higher switching current. In Figure 4.9b, the first unipolar J-E curve ($0 \rightarrow 2.5$ kV cm^{-1}) exhibits a bump at about 1.5 kV cm^{-1}, which must be due to the current induced by some dipole reversal since the sample was previously poled by a negative voltage. In the following J-E loops (3rd to 19th), no bump was observed in either the increasing-field or decreasing-field curve because most "dipoles" had been aligned in the positive direction during the first positive electric field application[14]. The J-E curves of MXene/P(VDF-TrFE-CFE) in Figure 4.9a and d exhibit typical relaxor ferroelectric behavior, which some previous reports have discussed[16-17]. We measured eight samples and found that they all exhibit similar behavior. In addition, the current density decreased and gradually approached a constant value when the field was cycled. In inset (i) of Figure 4.9b, the current density indicates a clearly decreasing trend as a function of the number of cycles at 2.5 kV cm^{-1}, which could be due to gradually decreasing dipole reversal under the positive unipolar electric field[18]. Relaxation measurements were performed to investigate the relaxation properties of the MXene/P(VDF-TrFE-CFE) composite. Inset (ii) of Figure 4.9b reveals that the current density experienced an exponential decay, which Equation 2 illustrates as follows:

$$\log[J(t)] = a\log(t) + b, \tag{2}$$

where $J(t)$ is the current density at time t, and a and b are constants. The substitution of several sets of original data into Equation 2 yielded the mean values of constants a (-0.276) and b (-0.1425). Therefore, Equation 2 can be written as follows:

$$J(t) = 0.72\,t^{-0.276}. \tag{3}$$

This long relaxation time observed in the MXene/P(VDF-TrFE-CFE) composite (Figure 4.9b) has significant implications for the mechanism of dielectric constant enhancement, which is discussed shortly.

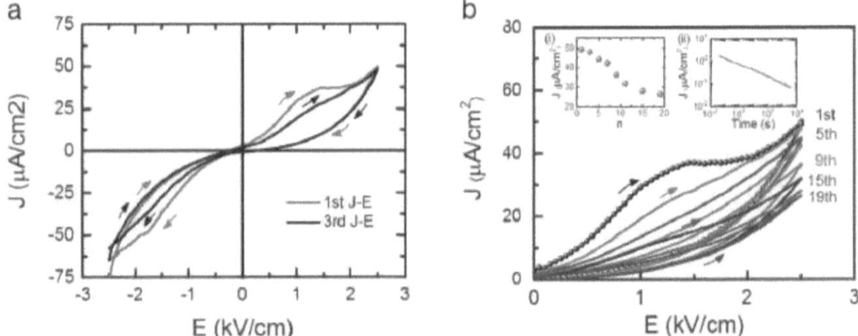

Figure 4.10 Composite: (a) The first *J–E* loop and third J-E curve of the MXene/P(VDF-TrFE-CFE) (8.0 wt % MXene) composite measured with an electric field of -2.5 kV cm^{-1} to 2.5 kV cm^{-1}; (d) The unipolar *J–E* loops measured with a positive field applied to the MXene/P(VDF-TrFE-CFE) (8.0 wt % MXene) composite, which has been pre-poled with a negative field of 2.5 kV cm^{-1}; Inset (i): Decrease in the current density at maximum field with the number of cycles (*n*) of the field; Inset (ii): Time-dependent leakage current density measured under a constant field of 1.25 kV cm^{-1} for one hour.

Figure 4.10a illustrates the leakage-corrected switching current density (left y-axis) and derived polarization (right y-axis) of MXene/P(VDF-TrFE-CFE) composites, both of which increase with increasing MXene loadings. The leakage-corrected polarization can be calculated as follows:

$$P(t) = \int_0^t \left[J(t) - J_d(t) \right] dt , \tag{4}$$

where *J(t)* is the measured total current density, including the current induced by dipole reversal and leakage, and *J$_d$(t)* is the measured current density after several cycles of electric field, which approximately represents the leakage current. Inset (i) of Figure 4.10a depicts the leakage-corrected hysteresis loop and leakage-corrected switching current density of the MXene/P(VDF-TrFE-CFE) composite with 8.0 wt % MXene loading. This result clearly indicates that some switching process takes place within the MXene-containing polymer, which is not the case in the pristine P(VDF-TrFE-CFE) polymer. The magnitude of the polarization ranged from 1.9 $\mu C/cm^2$ for a sample with 4.0 wt % MXene to 27.0 $\mu C/cm^2$ for a sample with 10.7 wt % MXene. These polarization values are relatively large for a polymer and indicate that some other dipole must have formed within the composite that does not exist in the pristine polymer.

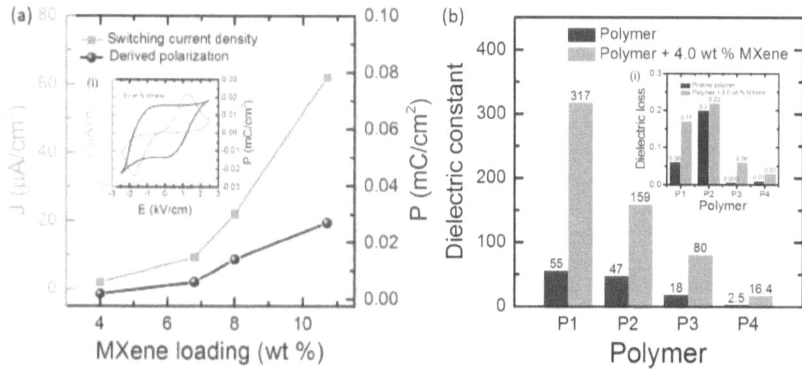

Figure 4.11 (a) The leakage-corrected switching current density and derived polarization of MXene/P(VDF-TrFE-CFE) composites with different MXene loadings; Inset (i): The leakage-corrected J-E and P-E loops of MXene/P(VDF-TrFE-CFE) with 8.0 wt % MXene loading; (b) The dielectric constant of different polymers without and with 4.0 wt % MXene; Inset (i): The dielectric loss of different polymers embedded with 4.0 wt % MXene loading; P1: P(VDF-TrFE-CFE), P2: P(VDF-TrFE-CTFE), P3: P(VDF-TrFE), P4: PVP.

Hence, we believe that the microscopic dipole model can more effectively describe the increased dielectric constant of these MXene-containing polymers. In this model, when an electric field (E_{ext}) is applied to the device (Figure 4.11), charges accumulate on the interfaces between conductive fillers (MXene sheets in our case) and the polymer matrix, where the microscopic dipoles are situated. This differs from the MXene-free polymer, where the polarization charge is accumulated near the electrodes of the MIM device (Figure 4.11a). In the microscopic dipole model, the MXene sheets do not need to be as close together as the microcapacitor model requires, which is the case for our samples. The microscopic dipole model can even explain the slow current relaxation that is apparent in Figure 4.10b, inset (ii). In true metallic fillers embedded in a polymer matrix, the relaxation of charges on the surface of metal fillers should occur rapidly. However, in our MXene/P(VDF-TrFE-CFE) composite, a long relaxation time was observed, which suggests that the positive and negative charges that accumulated on the interfaces between the MXene sheet and the polymer matrix will recombine slowly after the removal of the electric field. We believe that the long relaxation time is due to the fact that the MXene sheets, while metallic, have some polymer chains intercalated between them, as indicated by the increased interlayer spacing of MXene (from 1.2 nm to 1.4 nm). These polymer chains can slow down the recombination of the accumulated charges (at the interfaces between the MXene sheets and the polymer) after the removal of the external electric field.

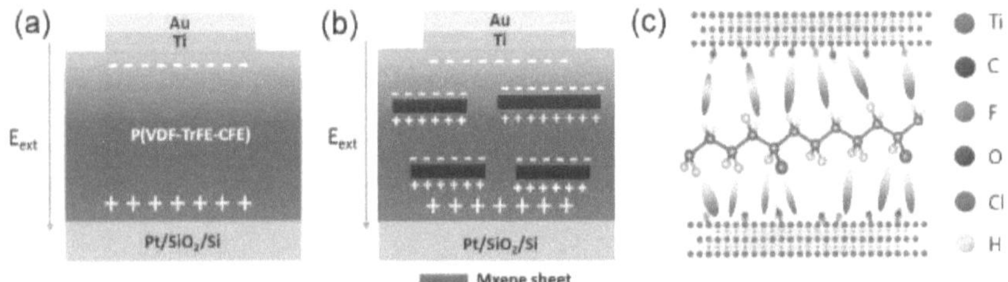

Figure 4.12 (a-b) Schematic illustration of the polarization charges that exist in pristine P(VDF-TrFE-CFE) and MXene/P(VDF-TrFE-CFE) MIM capacitors under an external electric field (E_{ext}); (c) Schematic illustration of dipoles that may form between MXene surface atoms (F, O) and H on the polymer backbone if polymer chains are intercalated between MXene sheets.

Table 4.5 The dielectric constant enhancement factor of MXene/polymer composites with MXene embedded in different polymer matrices.

	Weight percent	0	4.0
MXene/	Dielectric constant	55	317
P(VDF-TrFE-CFE)	Dielectric loss	0.06	0.17
	Enhancement factor	**1**	**5.7**
	Weight percent	0	4.0
MXene/	Dielectric constant	47	159
P(VDF-TrFE-CTFE)	Dielectric loss	0.2	0.22
	Enhancement factor	**1**	**3.4**
	Weight percent	0	4.3
MXene/	Dielectric constant	18	80
P(VDF-TrFE)	Dielectric loss	0.006	0.06
	Enhancement factor	**1**	**4.4**
	Weight pecrent	0	4.0
	Dielectric constant	2.5	16.4
MXene/PVP	Dielectric loss	0.01	0.03
	Enhancement factor	**1**	**6.6**

Finally, these MXenes-polymer composites may feature a special type of polarization deriving from hydrogen bonds that can form at any point where polymer chains manage to intercalate between the MXene sheets (see Figure 4.11c). As mentioned earlier, XRD data indicate that the MXene interlayer spacing expands at a higher polymer content, which suggests that some polymer chains may have successfully intercalated between MXene sheets. The hydrogen bonds between negative charges on the MXene surface (*e.g.* F and O) and H atoms on the polymer backbone can form a dipole that responds to the applied electric field. A previous report has proposed the large dipole moment of such hydrogen bond as the key factor for dielectric constant enhancement of a PVDF/hydrated metal salt composite. Moreover, another study has calculated the dielectric constant enhancement factor due to hydrogen bonds to be approximately three through a comparison of the dielectric constant of hydrogen-bonded and dipolar aprotic solvents[18]. Since the H-F and H-O bonds can exist in MXene-polymer composites, we surmise that hydrogen bonding between the atoms on the MXene surface and an H atom on the polymer backbone could contribute to the enhanced dielectric constant in our composites.

4.2.4   Conclusion

We have demonstrated that dispersing MXene in the P(VDF-TrFE-CFE) matrix induces a large enhancement of the dielectric constant, which can reach as high as 10^5 near the percolation limit of 15.3 wt % MXene loading. The ratio of the dielectric constant to loss tangent for MXene-loaded composites outperforms all previously reported conductive fillers incorporated in the same polymer. Current voltage and current relaxation measurements suggest that the origin of the substantial permittivity enhancement is primarily due to the microscopic dipoles formed by the accumulation of charges at the interfaces between the MXene fillers and the polymer matrix. This dielectric constant enhancement by MXene dispersion is also observed in other insulating polymers.

4.3 Size and Surface Termination Effect on Dielectric Properties

We report a strong effect of the MXene flake size and surface termination on the dielectric permittivity of MXene polymer composites. Specifically, poly(vinylidene fluoride-trifluoro-ethylene-chlorofluoroehylene) or P(VDF-TrFE-CFE) polymer embedded with large (ca. 4.5 μm) Ti3C2Tx flakes achieves a dielectric permittivity as high as 10^5 near the percolation limit of 15.3 wt % MXene loading. In comparison, the dielectric permittivity of MXene/P(VDF-TrFE-CFE) using small (ca. 1.5 μm) Ti3C2Tx flakes (S-MXene) achieves a dielectric permittivity of 10^4 near the percolation limit of 16.8 wt %. Meanwhile, increasing the concentration of surface functional groups on the MXene surface (−O, −F, and −OH) by extending the etching time gives a dielectric constant of 2204 near the percolation limit of 15.7 wt %. The ratio of permittivity to the loss factor of our large flake composite is superior to that of the small flake composite, and to all previously reported carbon-based fillers in P(VDF-TrFE-CFE). We show that the dielectric permittivity enhancement is strongly related to the charge accumulation at the surfaces between the two dimensional (2D) MXene flakes and the polymer matrix under an external applied electric field.

4.3.1 Size and Surface Chemistry Engineering

Figure 4.13a shows the AFM image of L-MXene flakes with a 5.0 μm lateral size. Figure 4.13b shows the AFM image of overlapping S-MXene sheets with a 1.5 μm lateral size. Figure 4.13c,d illustrate the size distribution of L-MXene and S-MXene, which reveal that the average lateral size of L-MXene and S-MXene is 4.5 and 1.5 μm, respectively. Additionally, XPS was used to characterize the surface chemistry of MXene sheets after 24 and 96 h etching. The analysis reveals the presence of −OH, −O, and −F functional groups on the Ti3C2Tx MXene surface. Compared with 24 h etching, the percentage of O (−O, −OH) and F increases from 14.5 and 10.2 to 20.0 and 14.3, respectively. These results indicate the oxidation of MXene flakes during the overetching process, from which we can conclude that the surface of MXene flakes after 96 h etching is rich in −F, −O, and −OH (Figure 4.13e).

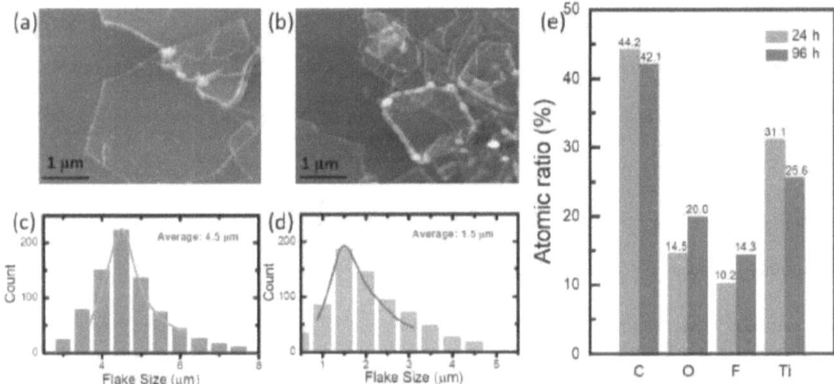

Figure 4.13 (a) AFM images of large-sized Ti3C2Tx MXene (L-MXene) flakes deposited on mica. (b) AFM images of small-sized Ti3C2Tx MXene (SMXene) flakes. (c) Statistical distribution of the L-MXene flake size. (d) Statistical distribution of the S-MXene flake size. (e) The concentration of surface termination on MXene sheets after 24 and 96 h etching, measured using XPS.

4.3.2 Dielectric Properties

Figure 4.14a displays the dielectric permittivity of L-MXene/P(VDF-TrFE-CFE), S-MXene/P(VDF-TrFE-CFE), and TL-MXene/P(VDF-TrFE-CFE) composites as a function of filler loading, which indicates that the dielectric permittivity of these composites initially increases with MXene filler concentration. A dielectric constant > 100000 was reached in the case of the composite with 15.3 wt % L-MXene in the terpolymer; in comparison, a dielectric constant of 10^5 was reached in the case of the S-MXene/polymer composite with a 16.8 wt % of S-MXene loading; the dielectric constant of the TL-MXene/terpolymer composite reached 2204 near the percolation limit at around 16.0 wt %. The obvious degradation of dielectric permittivity of TL-MXene/P(VDF-TrFE-CFE) composites can probably be ascribed to the high dense concentration of negative surface functional groups on the TL-MXene surface, which could e ffectively decrease charge accumulation on the TL-MXene/polymer interface under the externally applied electric field.

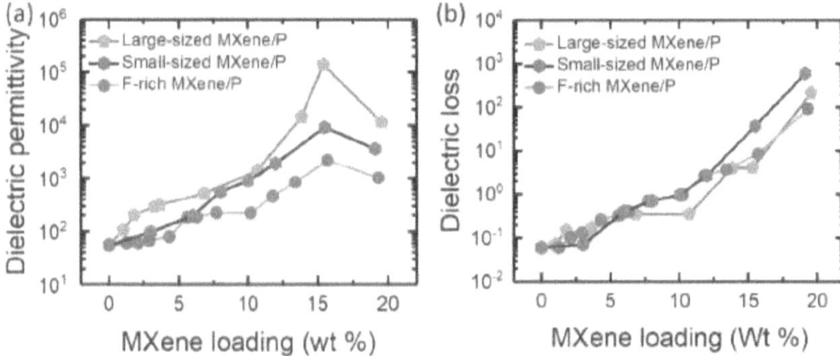

Figure 4.14 (a) Dielectric permittivity and (b) dielectric loss of the LMXene/P(VDF-TrFE-CFE), S-MXene/P(VDF-TrFE-CFE), and TLMXene/P(VDF-TrFE-CFE) composites versus MXene loading measured at room temperature and 1 kHz.MXene/polymer Fractional Order Capacitor.

4.3.3 Mechanism

We have developed a method to enhance the dielectric constant of MXene/P(VDF-TrFE-CFE) composites by engineering the flake size and concentration of surface functional groups. The MXene/P(VDF-TrFE-CFE) composite using large (4.5 μm) MXene flakes reaches a dielectric permittivity as high as 10^5 near the percolation limit. In comparison, the MXene/P(VDF-TrFE-CFE) composite using small MXene flakes reaches a dielectric permittivity of 10^4 near the percolation limit, while the dielectric constant of the TL-MXene/P(VDF-TrFE-CFE) composite reaches to 2204 near the percolation limit of 16.0 wt %. These results were explained based on the accumulation of charges at the interfaces between the two dimensional (2D) MXene flakes and the polymer matrix, where the switchable microscopic dipoles are situated.

4.3.4   Conclusion

We have developed a method to enhance the dielectric constant of MXene/P(VDF-TrFE-CFE) composites by engineering the flake size and concentration of surface functional groups. The MXene/P(VDF-TrFE-CFE) composite using large (4.5 μm) MXene flakes reaches a dielectric permittivity as high as 10^5 near the percolation limit. In comparison, the MXene/P(VDF-TrFE-CFE) composite using small MXene flakes reaches a dielectric permittivity of 10^4 near the percolation limit, while the dielectric constant of the TL-MXene/P(VDF-TrFE-CFE) composite reaches to 2204 near the percolation limit of 16.0 wt %. These results were explained based on the accumulation of charges at the interfaces between the two dimensional (2D) MXene flakes and the polymer matrix, where the switchable microscopic dipoles are situated.

4.4 Fractional Order Capacitor

Poly(vinylidene fluoride)-based polymers and their nanocomposites are used to fabricate electrostatic fractional-order capacitors. In this work, for the first time, two dimensional (2D) $Ti_3C_2T_x$ MXene/P(VDF-TrFE-CFE) composite is used to design/fabricate a fractional-order capacitor. The bandwidth of fabricated FOC is between 200 kHz-2 MHz, where the variation in phase angle is no more than $\pm 2°$. Additionally, the constant phase angle (CPA) of MXene/P(VDF-TrFE-CFE) can be precisely tuned from $-67°$ to $-34°$ by varying the volume ratio of MXene nanosheets in polymer matrix. The results presented in this work demonstrate the potential of the FOCs fabricated using MXene/P(VDF-TrFE-CFE) composites in realization of various electrical circuit systems.

4.4.1 Fabrication and Characterization

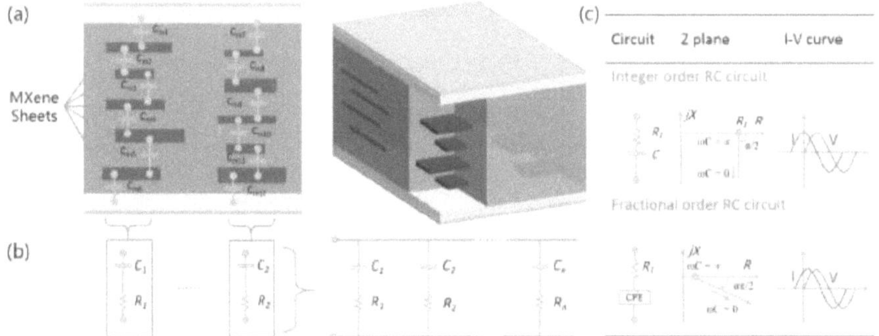

Figure 4.15 (a) A conceptual front view and a 3D schematic of the proposed fractional capacitor showing the top and bottom electrodes, the polymer matrix, and the MXene sheets. (b) the approxiamate equivalent network comprising an equivalent capacitance (C1, C2,......, etc.) in series with a resistor that models loss (R1, R2, ..., Rn). Horizontal fringing capacitances are ignored assuming that the thickness of the Mxene sheet is very small. (c) Impedance $Z(j\omega)$ loci of integer and fractional order RC circuits. CPE: constant phase element.

The presence of the conductive MXene sheets creates several capacitors that are connected in series (Cm1, Cm2,…), which can be equivalently represented as C1. Meanwhile, the dielectric loss can be modeled as a series resistor (R1). The overall capacitor can be modeled as a network very similar to a parallel-combination of series connected Rs and Cs, and the fractional impedance of the network can be tuned by varying the MXene loading (Figure 4.15). For integer RC circuit, $\alpha = 1$, $\angle Z(j\omega) = -90°$, and for fractional order one, $0 < \alpha < 1$, $-90° < \angle Z(j\omega) < 0°$.

In order to evaluate the performance of fabricated capacitors, we performed magnitude and phase angle of FOCs' impedance measurements with an Agilent 4980A LCR meter (Figure 4.16). We notice that the phase angle is near constant and stable from 200 kHz to 2 MHz, and the constant phase angle in this frequency range is increased by introducing the conductive filler $Ti_3C_2T_x$ MXene nanosheets into P(VDF-TrFE-CFE) matrix. This phenomenon can be explained by the interplay of Maxwell-Wagner-Sillers relaxation and dipolar relaxation.[19,20] We demonstrate a new class of FOCs that are fabricated using MXene/P(VDF)-based polymer percolated composites, which exhibits a tunable but stable CPA over a broad CPZ.

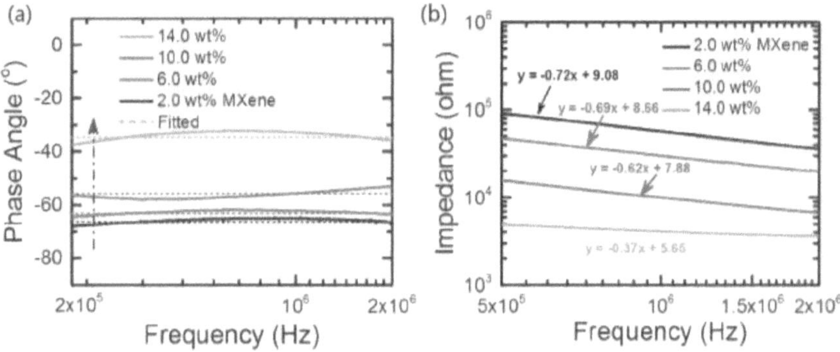

Figure 4.16 (a) The measured phase angle of impedance versus frequency from 200 kHz up to 2 MHz for different MXene loadings. (b) The impedance magnitude versus frequency in log scale.

4.4.2 Conclusion

In summary, we demonstrate solid-state fractional capacitors fabricated using MXene/P(VDF)-based polymer composites. The fractional capacitors exhibit a tunable but stable phase angle over a broad frequency range. The proposed compact capacitors possess dimensions in the micrometer range enabling integration with broad range of electronic devices. The percolated MXene-based composites are comprised of a large number of microcapacitors, whose phase angle can be changed by varying the loading of the MXene filler. Our MXene-based approach is an effective method for the realization of surface-mount fractional devices that will ultimately be used in commercial electronics.

REFERENCES

[1] M. N. Almadhoun, M. N. Hedhili, I. N. Odeh, P. Xavier, U. S. Bhansali, H. N. Alshareef. *Chem. Mater.* **2014,** *26*, 2856-2861.

[2] E. E. Shafee, M. E. Gamal, M. Isa. *J. Polym. Res.* **2012**, *19*, 9805.

[3] Z. Dan, J. Jiang, X. Zhang, J. Qian, Z. Shen, C. Nan, Y. Shen. Ceram. Int. 2020, 46, 1119-1123.

[4] T. Hu, J. Wang, H. Zhang, Z. Li, M. Hu, X. Wang. **2015**, *17*, 9997-10003.

[5] W. Lisowski, A. H. J. Berg, D. Leonard, H. J. Mathieu. *Surf. Interface Anal.* **2000**, *29*, 292–297.

[6] J. Halim, K. M. Cook, M. Naguib, P. Eklund, Y. Gogotsi, J. Rosen, M. W. Barsoum. *Appl. Surf. Sci.* **2016**, *362*, 406-417.

[7] C. W. Nan, Y. Shen, J. Ma. *Annu. Rev. Mater. Res.* **2010**, *40*, 131-151.

[8] S. H. Zhang, N. Y. Zhang, C. Huang, K. L. Ren, Q. M. Zhang. *Adv. Mater.* **2005**, 17, 1897-1901.

[9] J. W. Wang, Q. D. Shen, H. M. Bao, C. Z. Yang. Macromolecules 2007, 38, 2247-2252.

[10] C. Huang, Q. M. Zhang *Adv. Funct. Mater.* **2004**, *14*, 501-506.

[11] A.Javadi, Y. Xiao, W. Xu, S. Gong. *J. Mater. Chem.* **2012**, *22*, 830-834.

[12] C. W. Nan, Y. Shen, J. Ma. *Annu. Rev. Mater. Res.* **2010**, *40*, 131-151.

[13] M. Zhu, Y. Huang, Q. Deng, J. Zhou, Z. Pei, Q. Xue, Y. Huang, Z. Wang, H. Li, Q. Huang, C. Zhi. *Adv. Energy Mater.* **2016**, *6*, 1600969.

[14] B. Dickens, E. Balizer, A. S. DeReggi, S. C. Roth. *Journal of Applied Physics* **1992,** *72*, 4258-4264.

[15] H. M. Bao, J. F. Song, J. Zhang, Q. D. Shen, C. Z. Yang. *Macromolecules* **2007**, *40*, 2371-2379.

[16] L. Zhu. *J. Phys. Chem. Lett.* **2014**, *5*, 3677-3687.

[17] Prateek, V. K. Thakur, R. K. Gupta. *Chem. Rev.* **2016**, *116*, 4260-4317..

[18] Y. Yao, Q. Wang, H. Wang, B. Zhang, C. Zhao, Z. Wang, Z. Xu, Y. Wu, W. Huang, P. Y. Qian, X. X. Zhang. *Adv. Mater.* **2013**, *25*, 711-718.

[19] A. Agambayev, S. P. Patole, M. Farhat, A. Elwakil, H. Bagci, K. N. Salama, *Chem. Electro. Chem.* **2017**, *4*, 2807-2813.

[20] G. Zhang, D. Brannum, D. Dong, L. Tang, E. Allahyarov, S. Tang, K. Kodweis, J. K. Lee, L. Zhu, *Chem. Mater.* **2016**, *28*, 4646-4660.

Chapter 5- Ti$_3$C$_2$T$_x$ MXene/in-plane Aligned PVDF Actuator

MXene/polymer bimorph are utilized to fabricate an efficient flexible solar tracking system based on a photo-thermo-mechanical (PTM) actuator operated under uniaxial deformation. The actuation function of the proposed device originates from integrating surface plasmon-assisted photo-thermal MXene with thermo-mechanical 3D aligned polyvinylidene fluoride (PVDF). Two types of tracking options are simulated based on the experimental PTM actuator deformation data. We find that uniaxial East-West option improves the overall energy intensity of the solar module by over 30 %, in comparison with the optimized tilting-fixed option. We also demonstrate the thermally driven self-oscillation of MXene/PVDF actuator, which may have promising potential for nanogenerators. In addition, PTM actuator devices display robust mechanical strength and durability, with no noticeable degradation of their performance after > 500 cycles.

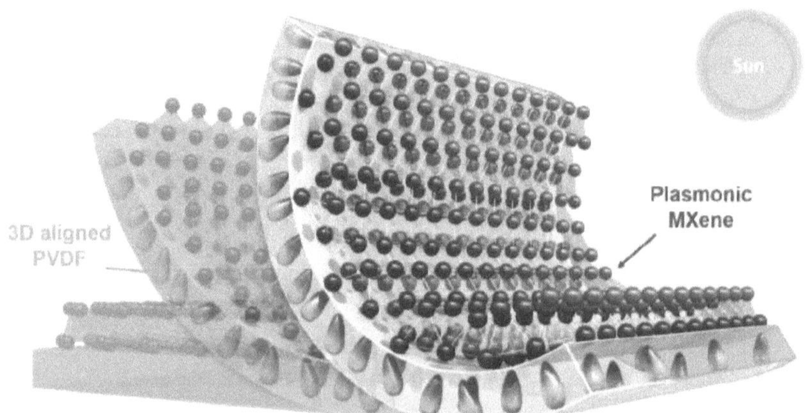

Figure 5.1 Schematic illustration of the photo-thermal-mechanical (PTM) actuator based on MXene/3D aligned PVDF.

Figure 5.1 illustrates the design concept of our photo-thermo-mechanical (PTM) actuator based on 2D MXene/three dimensional (3D) aligned PVDF. Flexible photovoltaic (PV) modules represent an attractive option for energy conversion and storage applications in wearable self-charging power systems[1-3]. Indeed, conventional fixed PV modules suffer from low annual energy efficiency and power output due to their average high incident angle between sunlight and solar cells over the daytime. In order to minimize their incident angle, the concept of sunlight tracking was introduced to the PV industry, where one or two electric mechanical tilting axes are usually utilized according to the geographic location of the modules[4]. Although electro-mechanical tracking systems can successfully provide over 20% increase in annual energy conversion compared with non-tracking solar modules, such tracking systems are not widely used in industry due to their costly installation[5-6]. Because of their complex installation process and large size, conventional electro-mechanical tracking systems are costly and not suitable for flexible wearable energy storage devices. One of the autonomous solar tracking systems that has recently generated a lot of interest from the research community is the bioinspired actuator, due to its high durability, lightweight wearability, and multi-stimulus responsiveness[7-10]. By integrating large thermal expansion polymer into surface plasmon-enhanced photo-thermal two-dimensional materials, built-in photo-thermo-mechanical (PTM) actuators can achieve high levels of sensitivity to sunlight stimulus, which make them ideal candidates for flexible solar tracking systems.

Three dimensional aligned PVDF exhibits extremely large thermal expansion along specific directions, which makes it a suitable polymer for controllable thermally-driven actuator applications. The photo-thermal effect is essential for the responsiveness of actuators to light stimulation, which can be enriched by the collective oscillations of free charge carriers, commonly known as surface plasmons (SPs)[11-13]. Several plasmonic two dimensional materials have been already exploited, such as graphene[14], black phosphorus[15], hexagonal boron nitride[16] and topological insulators[17]. MXenes, a new class of 2D-materials with the general formula $M_{n+1}X_nT_x$, in which M is an early transition metal, X is a carbon and/or nitrogen, T_x stands for the surface functional groups (e.g. OH, O, and/or F groups). In particular $Ti_3C_2T_x$ has also been proved to be plasmonic[18-21], which has shown great potential in the applications of photo-thermal therapy[22], surface-enhanced Raman spectroscopy (SERS) [23], photodetectors[24] and photonic diodes[25].

The study presented in this paper demonstrates for the first time that integrating a surface plasmon enhanced photo-thermal MXene with a thermo-mechanical 3D aligned PVDF to fabricate actuators. These actuators can work as highly efficient solar tracking devices for flexible solar module by using its in-built photo-thermo-mechanical uniaxial deformation. In this study, we simulated and found that it is able to obtain over 30 % energy conversion improvement using the PTM actuator tracking systems compared to the tilting-fixed system. We showed the thermo-mechanical self-oscillation of PTM actuators films on a hot plate, which can be applied to nanogenerators.[26-27].

5.1 Fabrication and Characterization

For thermo-mechanical bilayer actuator, an obvious mismatch in the coefficient of thermal expansion (CTE) between active and coated materials is required to enable an effective response to an external thermal stimulation. PVDF is suitable as an active layer for a thermal actuator, due to its extremely high CTE value and excellent mechanical flexibility. The 3D aligned PVDF film has an anisotropic CTE, as can be seen from the visible transverse thermal expansion ($\sim$ 30 μm) and negligible thermal expansion along the longitudinal direction ($\sim$ 0) when heated from 23 $^{\circ}$C to 60 $^{\circ}$C (Figure 5.2).

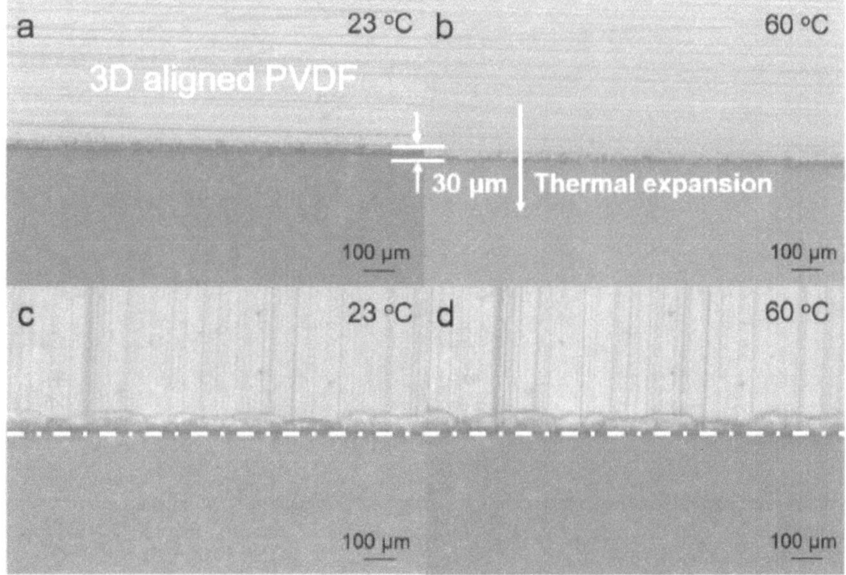

Figure 5.2 (a-b) Thermal expansion of 3D poled PVDF film along the transversal direction from 23 °C to 60 °C. (c-d) Thermal expansion of 3D poled PVDF film along the longitudinal direction over the same temperature range.

Figure 5.2a-b present the surface scanning electron microscopy (SEM) and atomic force microscopy (AFM) images of our 3D aligned PVDF film, showing its 3D alignment along the longitudinal direction. The characteristic peak at $\theta = 20.6^0$ of X-ray diffraction (XRD) and the Fourier-transform infrared spectroscopy (FTIR) analysis respectively reveal the existence of a predominant β phase[28] and corresponding vibrational modes[28] (Figure 5.2c-d). The anisotropic expansion of 3D aligned PVDF makes it possible to control the direction of the deformation of the actuator.

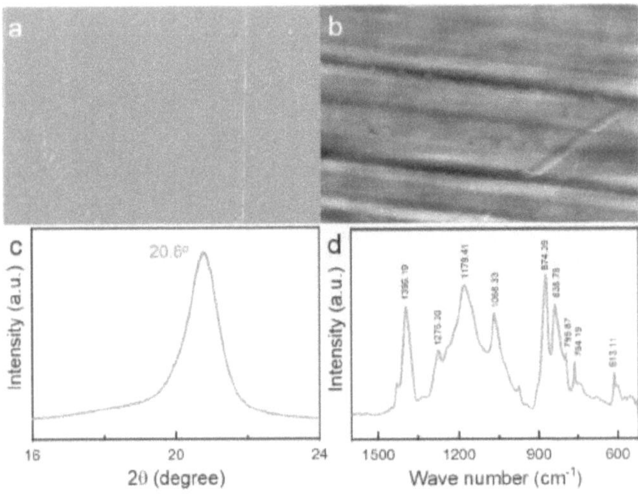

Figure 5.3 (a) Scanning electron microscopy (SEM) image, (b) Atomic force microscopy image, (c) X-ray diffraction (XRD) pattern, and (d) Fourier transform infrared (FT-IR) spectra of 3D poled PVDF film.

$Ti_3C_2T_x$ MXene is a good candidate as passive layer in the photo-thermo-mechanical actuator due to its low CTE and surface plasmon and photothermal efficiency. Figure 5.4 shows that there was almost no thermal expansion of Ti3C2Tx when heated from 23 ^{0}C to 60 ^{0}C.

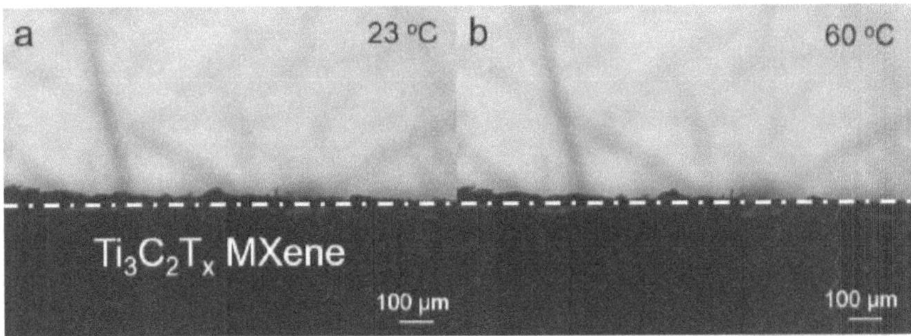

Figure 5.4 (a-b) Thermal expansion of $Ti_3C_2T_x$ MXene film from 23 ^{0}C to 60 ^{0}C.

Figure 5.5a-b shows the morphology and XRD of $Ti_3C_2T_x$ MXene flakes. The X-ray photoelectron spectra (XPS) indicate the negative functional groups on the $Ti_3C_2T_x$ MXene surface[29] (Figure 5.5c-d), which guarantees its strong adhesion on the plasma treated hydrophilic PVDF surface, thus ensuring the mechanical robustness of PTM actuator to bear substantial and repeated deformations.

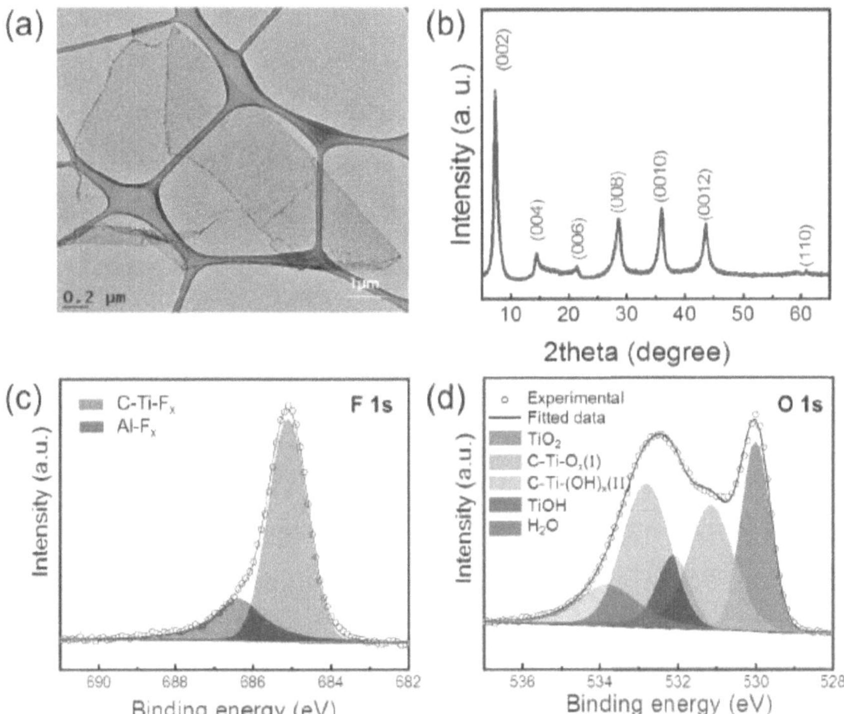

Figure 5.5 (a) Transmission electron microscopy (TEM) image. (b) XRD pattern. (c) F 1s, and (d) O 1s core levels of $Ti_3C_2T_x$, as obtained by X-ray photoelectron spectroscopy (XPS).

Figure 5.6a shows the typical absorption spectrum (200-1000 nm) of $Ti_3C_2T_x$, with a two characteristic spectral bands at ca. 340 and 760 nm. The former is assigned to the interband transitions of $Ti_3C_2T_x$ (*19-20*), while the latter band is ascribed to the LSPR effect at the edge of the visible-NIR spectral regime. The plasmonic behavior was further corroborated by means of scanning transmission electron microscopy (STEM) combined with electron energy loss spectroscopy (EELS). Figure 5.6b depicts the annular dark field (ADF) image of a triangular $Ti_3C_2T_x$ flake supported by a lacey carbon film, in addition to the corresponding EELS-fitted intensity maps of various surface plasmon (SP) modes and interband transitions (with an onset at 3.5 eV) in $Ti_3C_2T_x$, which were distinctly visualized as described in (*20, 21*). The unique spatial distribution of each of the detected SP modes is dependent on the oscillation frequency of the free electron gas, either along (0.32, 0.54 and 0.8 eV) or perpendicular to (1.7 eV) the main axes of the flakes. Interestingly, the energy of the SP mode visualized at ca. 1.7 eV correlates very well with the absorption peak at ca. 760 nm (Figure 5.6a), confirming that this peak is due to an LSPR effect in $Ti_3C_2T_x$. Besides, owing to the homogeneous spatial distribution of that particular SP mode, which does not depend on the geometry of the flake, we believe that a film made of several (ca. hundreds) multi-stacked $Ti_3C_2T_x$ flakes would exhibit the same plasmonic feature (1.7 eV) obtained on a single flake. The abovementioned findings shows that the exceptional sensitivity of the fabricated PTM actuators is enhanced by the surface plasmon-assisted and photo-thermal properties of $Ti_3C_2T_x$ along with the orientable, extremely large coefficient of thermal expansion (CTE) mismatch with of the aligned PVDF polymer film. We note that even though PVDF polymer is known to exhibit pyroelectricity and piezoelectricity, we do not believe that the observed actuation in our devices is caused by either the pyroelectric or piezoelectric effects. The photo-thermo-mechanical expansion of 3D aligned PVDF film is negligible due to the quite low gradient of temperature variation versus time in ambient circumstance[35].

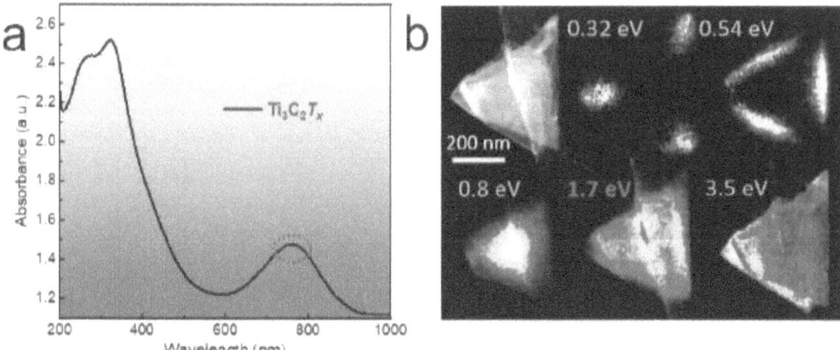

Figure 5.6 (a) UV-visible-NIR absorption spectrum of Ti$_3$C$_2$T$_x$ MXene (dispersed in deionized water). The red circle denotes the LSPR-dependent absorption peak at ca. 760 nm. (b) STEM-ADF image of a Ti$_3$C$_2$T$_x$ flake, along with the corresponding EELS intensity maps of its SP modes (0.32, 0.54, 0.8 and 1.7 eV) and interband transitions (3.5 eV). The boundaries of the multi-stacked Ti$_3$C$_2$T$_x$ flake is defined by a white dotted contour.

. Figure 5.7 illustrates the fabrication process of Ti$_3$C$_2$T$_x$ MXene/3D aligned PVDF PTM actuator. The process involves a hydrophilic treatment of the surface, followed by sequentially drop casting PEDOT:PSS and MXene layers, and finally cutting the device in the desired size and orientation. 80 μm thick 3D aligned polyvinylidene difluoride (PVDF) thin film was purchased from Fils Co., Ltd company. PEDOT:PSS aqueous solution was drop cast on the both side of plasma treated 3D aligned hydrophilic PVDF thin film and dried at room temperature overnight. Then, the Ti$_3$C$_2$T$_x$ MXene ink was cast on the PVDF/ PEDOT:PSS and dried at room temperature. After repeating drop-casting MXene ink for several times, the PTMA film with desired thickness was cut along different directions as needed. Notably, PEDOT:PSS layer with the same thickness was cast on the both side of 3D aligned PVDF substrate to increase the adhesion between Ti$_3$C$_2$T$_x$ MXene and 3D aligned PVDF.

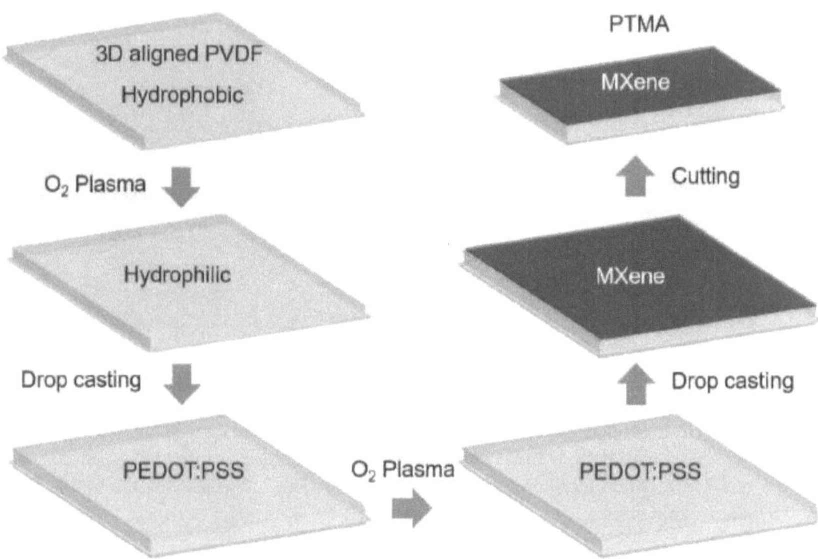

Figure 5.7 Fabricating process of MXene/PEDOT:PSS/3D aligned PVDF PTMA.

A PTM actuator up to 100 cm^2 was prepared in order to easily obtain several PTM actuator strips cut at different alignment angles, by carefully choosing the cutting direction (Figure 5.8a). The multi-layered microstructure of fabricated PTM actuator was further confirmed by its cross-sectional SEM image (Figure 5.8b).

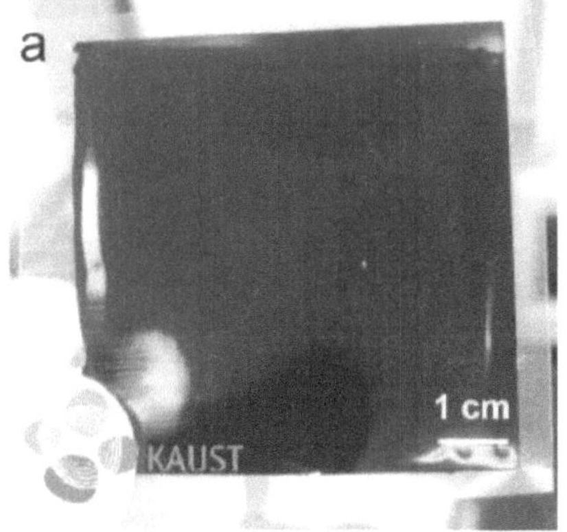
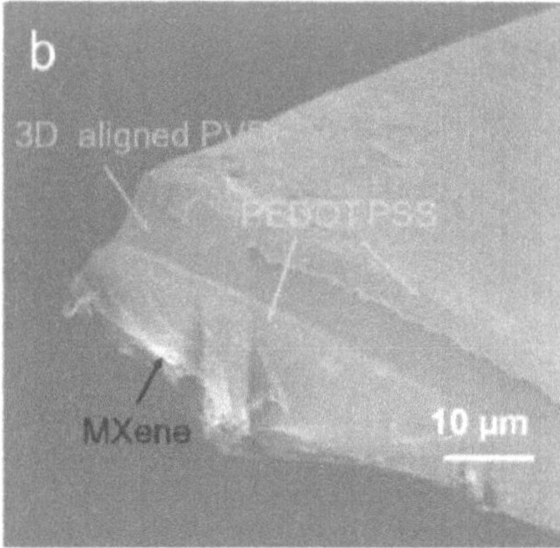

Figure 5.8 (a) An optical image and (b) cross-sectional Scanning Electron Microscopy image of PTMA. The scale in b and c are 1 cm and 10 µm, respectively.

5.2 Thermo-mechanical Locomotion

The precisely aligned PTM actuator experienced a quick bending deformation on a hot surface with $Ti_3C_2T_x$ MXene layer facing upwards, which can be attributed to the thermal expansion mismatch between the aligned PVDF and $Ti_3C_2T_x$ MXene layers. Figure 5.9a shows the infrared images of a 90^0 cut PTM actuator (1 x 5 cm) placed on a 55 °C hot plate. The figure shows time progression of the radius and curvature variation and thermal distribution along the longitudinal direction of the PTM actuator heated at different temperatures. The circle-fitted bending radius of the PTM actuator decreases from 2.4 cm to 0.6 cm while the temperature of PTM actuator increases from 39.3 °C to 55.2 °C, meaning that the corresponding curvature increases from 0.42 to 1.67 cm^{-1} within the same temperature range (Figure 5.9b). It is worth noting that the deformation of PTM actuator can be observed even at a temperature of 26.5 °C, which indicates a high sensitivity to external thermal stimulation.

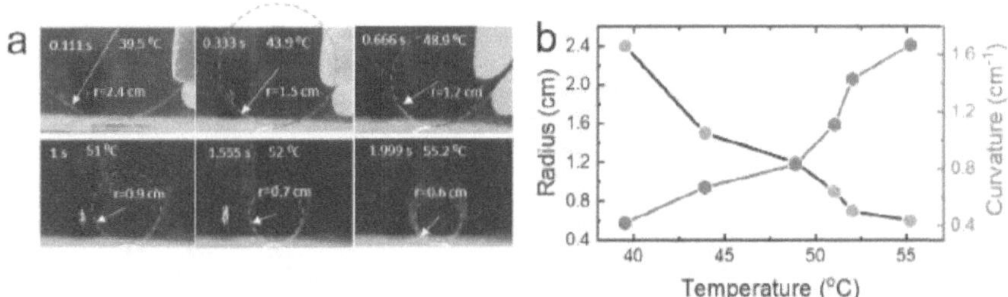

Figure 5.9 (a) Infrared images of the actuator at the beginning seconds on a 55 °C hot surface. (b) Radius and curvature of the actuator corresponding to (a).

Figure 5.10 illustrates the thermo-mechanical oscillation of 90° cutting PTMA on a hot surface, associated with a cyclical shift of the gravity center, due to its thermal exchange with air and hot surface. After undergoing a fast deformation in the first few seconds, we found that the PTMA lost its balance between two wings due to an uneven thermal expansion caused by the disturbance of the air-PTMA heat exchange. When the PTMA shifted to its right side, the right wing bended as temperature increased on the hot plate and the left wing expanded as temperature decreased under cold air, which led to a left shift of the gravity center and pulling the PTMA back to its left side. The inertia of the locomotion and continuous non-equivalent heat exchange of two wings formed a dynamic equilibrium with the gravity damping of PTMA, and finally exhibited a microscopic thermal driven oscillation within a certain temperature range.

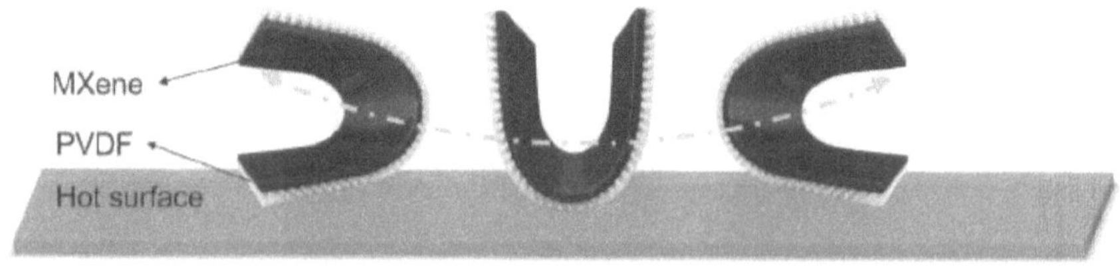

Figure 5.10 Schematic illustration of the oscillation of PTM actuator initiated by heat exchange.

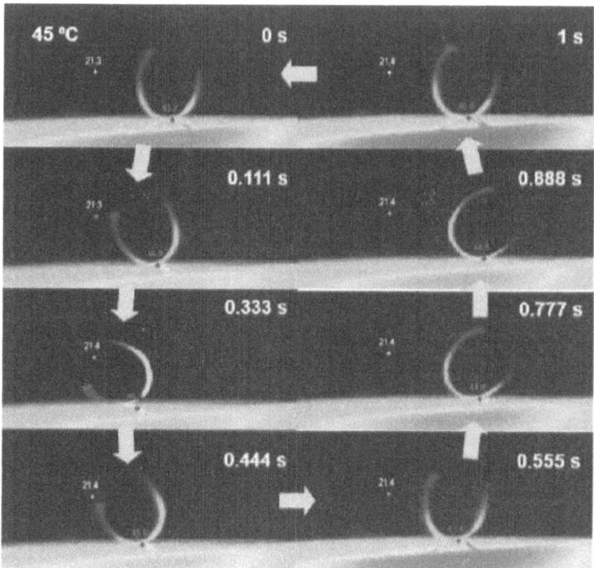

Figure 5.11 Time-dependent infrared images of a complete oscillation of the actuator.

Figure 5.11 shows in detail thermally-driven oscillations of a PTMA film placed on a 45 °C hot plate, with the air temperature constant at 21.4 °C. After achieving a dynamic equilibrium oscillation, it took 0.333 s for the PTMA film to swing from its initial position to its leftmost position, which was followed by swinging back to the starting location, then reaching to the rightmost position at 0.777 s, and finally returning to the initial position at 1 s. The infrared images show the temperature distribution of PTMA at each step in the oscillation process, and the variations of temperature clearly indicate the existence of a heat exchange between the PTMA and hot plate and cold air, respectively. Notably, the critical temperature of cyclic thermo-mechanical motion of PTMA (45 °C) is much lower than that of previously reported PDG-CNT/PVDF actuator

(55 °C)[30-31]. The ultrasensitive thermo-mechanical properties of PTMA show great potential for applications such as low grade waste energy conversion, thermal stimulated actuator, and soft robotic materials[32-34]. Here, we focus our research for applications in soft solar tracking systems.

In addition, we demonstrate that the PTM actuators display a robust mechanical strength and durability, with no noticeable degradation in performance after over 500 cycles (Figure 5.12).

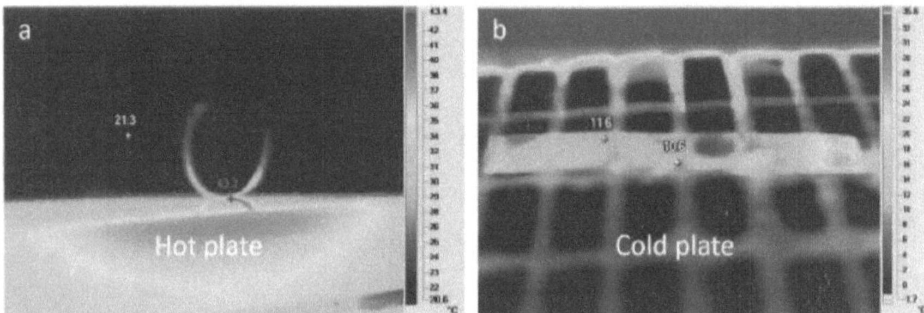

Figure 5.12 (a) Thermo-mechanical deformation of PTM actuator on a 45 °C hot plate. (b) Recover of bended PTM actuator on a cold surface (durability > 500 times).

5.3 Photo-thermo-mechanical Actuator

In order to fully understand the photo-thermo-mechanical properties of the $Ti_3C_2T_x$ MXene/3D aligned PVDF structure, we investigated the deformation of PTM actuator cut along 90°, 45°, and 0° (1×5 cm) under sunlight irradiation (Figure 5.13a-c). We found that the tilting orientation of PTMA was always perpendicular to the aligned direction of the PVDF substrate when applying an external thermal stimulation. Typically, the 90°-cut and 0°-cut of PTMA (visible by naked eyes) resulted in normal and parallel uniaxial bending under the sunlight, respectively. This was the case even for an ambient temperature as low as 28 °C (Figure 4.13a and c). Under the same conditions, the 45°-cut PTMA underwent anti-clockwise biaxial chiral distortion (Figure 5.13b).

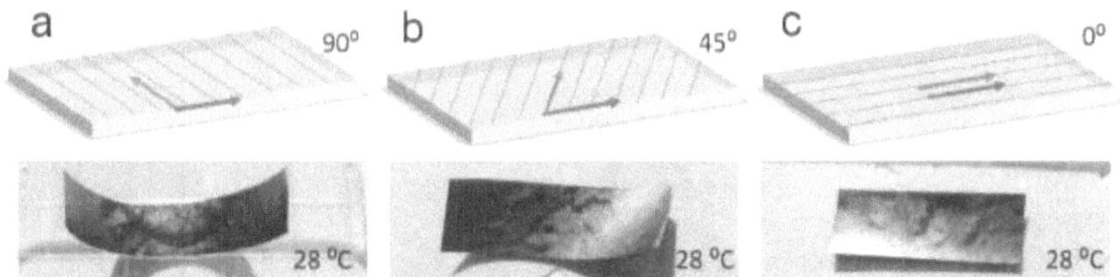

Figure 5.13 (a-c) Schematic illustration of PTMA films cut along different directions ($\varphi = 90°$, 45°, 0° and their corresponding deformation under sunlight.

For comparison, we investigated the photo-thermal-mechanical response of pristine MXene free standing film and PVDF substrate individually, neither of them presented deformation under sunlight (Figure 5.14).

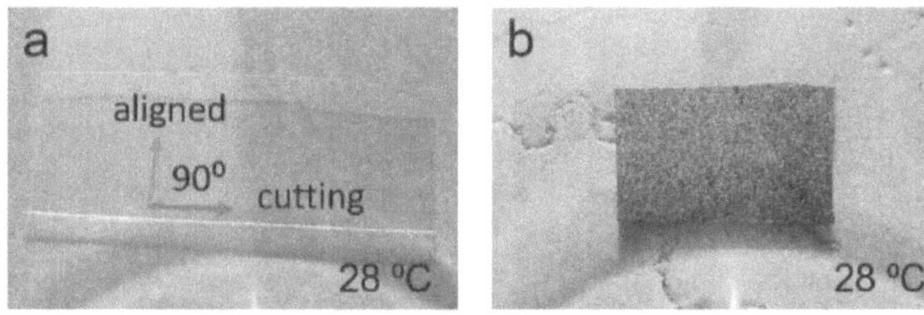

Figure 5.14 Optical images of (a) 90°-cut PVDF and (b) $Ti_3C_2T_x$ MXene free standing film under sunlight.

5.4 PTM Actuator for Solar Tracking

One promising application of this self-distorting PTMA in the photovoltaic (PV) industry is for active solar tracking systems that do not require electricity to drive. Figure 5.15a-b define the sun position across the sky, solar zenith angle (θ) and solar azimuth angle (φ) change with the earth's revolution and rotation. To maximize the solar illumination, PV modules are normally installed with a tilt angle β (the angle between the plane of the module and the horizontal surface) and an azimuth angle γ (the angle between the plane of the module and due north). For the tilting-fixed control module[36-37], β is set to be equal to the lattice, γ equals 180° for the Northern hemisphere and 0° for the Southern hemisphere. Although the fixed module saves the installation cost, it results in a low effective use of the solar energy due to the limited time within a day for a small-angle incidence (ϕ). In order to improve the efficiency of module, the concept of solar tracking is introduced, with the module dynamically rotating to follow the position of the sun. Theoretically, if there is no limitation on the module rotation, the ideal solar tracking is to maintain the relation of $\beta = \theta$ and $\gamma = \varphi$ all the time, which means the module is constantly perpendicular to the sunlight, enabling the maximum solar illumination on to the PV panel ($\phi = 0°$). The tracking system, via uniaxial tilting, can partially or fully achieve this purpose. Conventionally, a tracking system includes an electrically driven mechanical set-up and a rotating feedback system, both requiring heavy costs of installation, operation and maintenance. Thus, it is highly advantageous to study the self-tilting photo-thermo-mechanical actuator (PTMA) for a solar tracking system.

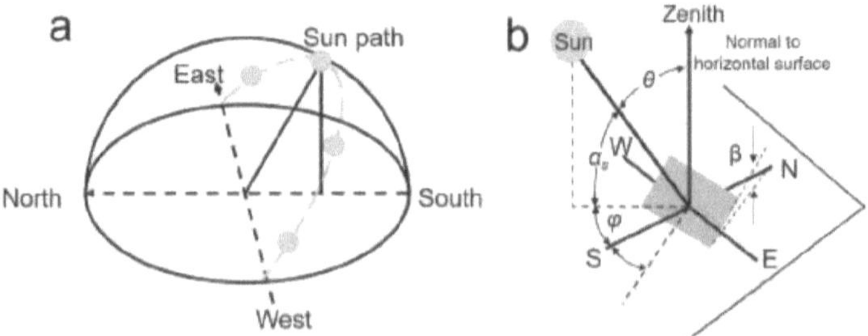

Figure 5.15 Schematic illustration of a) the sun path across the sky, b) the zenith angle and azimuth angle.

To demonstrate and evaluate the gain of the solar tracking system based on the PTMA, the solar spectrum calculator from the PV lighthouse was utilized to simulate the solar spectrum and solar incident angle within a specific day (17 Sep 2019, the tilting angle of PTMA was experimentally measured outdoor) at KAUST (22.3095° N, 39.1047° E). Additionally, the PV module efficiency and power output were obtained by using the light ray tracking simulation and built-in equivalent circuit model in sunsolver of PV lighthouse. Figure 5.16a-b illustrate the solar tracking modules of uniaxial North-South (N-S) tilting and uniaxial East-West (E-W) tilting, respectively. When including the tilting-fixed control module, a total of three tracking options were considered in our discussion.

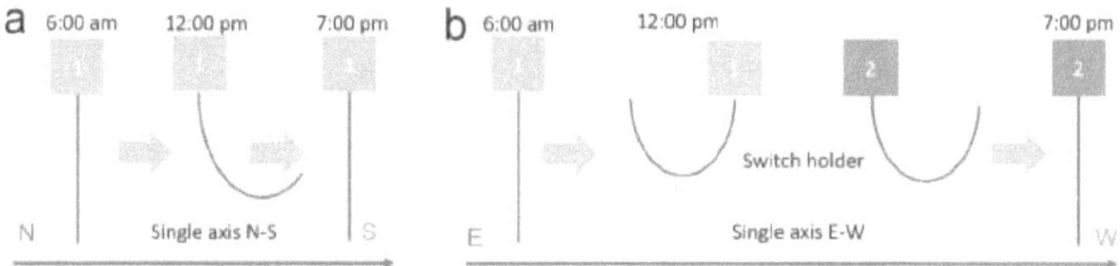

Figure 5.16 (a) uniaxial north-south (N-S) tilting, (b) uniaxial east-west (E-W) tilting. The square boxes in d and e represent holders.

Figure 5.17 Tilting angle evolution of $Ti_3C_2T_x$ MXene/3D aligned PVDF actuator over time due to its Photo-thermo-mechanical deformation in uniaxial N-S option.

Figure 5.18 Tilting angle evolution of $Ti_3C_2T_x$ MXene/3D aligned PVDF actuator over time due to its Photo-thermo-mechanical deformation in uniaxial E-W option.

Figure 5.19a-b plot the module orientation for all three options and the solar position. For the solar zenith angle, the uniaxial N-S offers better tracking than the uniaxial E-W option. In terms of the azimuth angle, the uniaxial E-W option showed optimized tracking results (Figure 5.17-5.18). A combination of the zenith and azimuth angles, and the incident angle ϕ of the three mentioned options are depicted in Figure 5.19c[38]. The difference of the incident angle between the control option and the uniaxial N-S option is negligible, due to the absence of tracking effect in the solar azimuth angle, even though it shows much better tracking in the solar zenith angle. As expected from the azimuth angle tracking, the uniaxial E-W option show optimized incident angle tracking. The direct and the diffuse solar spectrum on the plane of the module and PV efficiency can be calculated based on the incident angle. Typically, the smaller the incident angle, the more solar energy reaches the PV module plane[39].

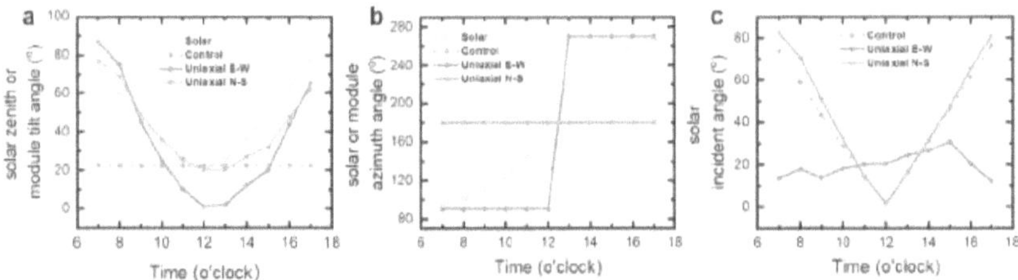

Figure 5.19 (a) Solar zenith θ or module tilt angle β. (b) Solar azimuth angle φ or module azimuth angle γ. (c) Solar incident angle φ.

5.5 PV Simulation

In this work, a light ray tracing simulation and equivalent circuit model were used to predict the PV efficiency and the PV output power. During the simulation, both direct and diffused solar spectra were calculated based on the incident solar angle. Figure 5.20a and b respectively display the simulated PV efficiency and the PV output power intensity of the control option and uniaxial E-W option. The uniaxial N-S option was omitted in this simulation, considering that its incident angle was quite close to that of the control option. Compared to the control option, the PV efficiency of the uniaxial E-W option resulted in a significant improvement during the early morning (6:00 am-10:00 am) and late afternoon (2:00 pm-6:00 pm) periods (Figure 5.20a).

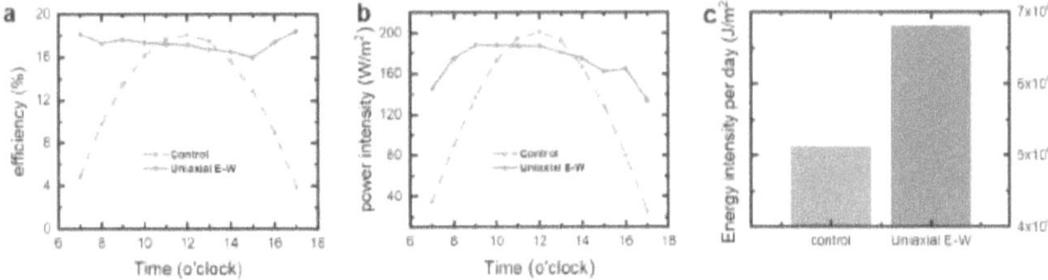

Figure 5.20 (a) Efficiency and (b) power intensity of module within the daytime. (c) Energy intensity per day of different module groups.

The output power of the module, a function of both the PV efficiency and solar spectrum, displayed an upward bended shape due to the fact that the solar intensity maximized around noon time and lowers down in the early morning and late afternoon (Figure 4.20b). Notably, the uniaxial E-W modules present higher overall power intensity than the control group. Figure 5.20c depicts the energy intensity of the given day by integrating the powder intensity with time, which is assumed to be constant in every hour. The uniaxial E-W solar tracking system based on PTM actuator improved the energy output by over 30% compared to the tilting-fixed control PV system.

5.6 Conclusion

In summary, we have demonstrated the solar tracking system based on the photo-thermo-mechanical (PTM) deformation of plasmonic $Ti_3C_2T_x$ MXene/3D aligned PVDF actuator. The sensitivity of the PTM actuator to photo stimulation is due to the marked enhancement of the surface plasmon resonance of $Ti_3C_2T_x$ MXene. In this work, three types of tracking options were simulated, where uniaxial E-W options improved the overall energy intensity of solar module by over 30 % compared to the optimized tilting-control option. Our results clearly show an improvement in PV efficiency, PV power intensity, and energy intensity based on our MXene/PVDF PTM actuator. Our design provides a new way to achieve solar tracking function using photo-thermal-mechanical property of two dimensional materials.

REFERENCES

[1]   A. Hauch, A. Georg, U. O. Krasovec, B. Orel, *J. Electrochem. Soc.* **2002**, *149*, A1208-A1211.

[2]   X. Y. Xue, S. H. Wang, W. X. Guo, Y. Zhang, Z. L. Wang, *Nano Lett.* **2012**, *12*, 5048-5054.

[3]   X. Pu, L. Li, H. Song, C. Du, Z. Zhao, C. Jiang, G. Cao, W. Hu, Z. L. Wang, *Adv. Mater.* **2015**, *27*, 2472-2478.

[4]   A. Lamoureux, K. Lee, M. Shlian, S. R. Forrest, M. Shtein, *Nat. Commun.* **2015**, *6*, 8092.

[5]   S. Bhaduri, L. M. Murphy, Wind Loading on Solar Collectors (**1985**).

[6]   GE Energy. Western Wind and Solar Integration Study. NREL/SR-550-47434 (**2010**).

[7]   L. Hines, K. Petersen, G. Z. Lum, M. Sitti, *Adv. Mater.* **2017**, *29*, 1603483.

[8]   N. W. Bartlett, M. T. Tolley, J. T. B. Overvelde, J. C. Weaver, B. Mosadegh, K. Bertoldi, G. M. Whitesides, R. J. Wood, *Science* **2015**, *349*, 161–165.

[9]   M. Wehner, R. L. Truby, D. J. Fitzgerald, B. Mosadegh, G. M. Whitesides, J. A. Lewis, R. J. Wood, *Nature* **2016**, *536*, 451-455.

[10]   W. Hu, G. Z. Lum, M. Mastrangeli, M. Sitti, *Nature* **2018**, *554*, 81-85.

[11]   J. Nelayah, M. Kociak, O. Stephan, F. J. G. de Abajo, M. Tencé, L. Hernard, D. Taverna, I. Pastoriza-Santos, L. M. Liz-Marzan, C. Colliex, *Nat. Phys.* **2007**, *3*, 348–353.

[12]   F.-P. Schmidt, H. Ditlbacher, U. Hohenester, A. Hohenau, F. Hofer, J. R. Krenn, *Nat. Commun.* **2014**, *5*, 3604.

[13]   J. M. Luther, P. K. Jain, T. Ewers, A. P. Alivisatos, *Nat. Mater.* **2011**, *10*, 361–366.

[14]   T. Low, P. Avouris, *ACS Nano* **2014**, *8*, 1086–1101.

[15]   M. A. Huber, F. Mooshammer, M. Plankl, L. Viti, F. Sandner, L. Z. Kastner, T. Frank, J. Fabian, M. S. Vitiello, T. L. Cocker, R. R. Huber, *Nat. Nanotechnol.* **2017**, *12*, 207–211.

[16]   C. T. Pan, R. R. Nair, U. Q. Bangert, R. R. Jalil, R. Zan, C. R. Seabourne, A. J. Scott, *Phys. Rev. B: Condens. Matter Mater. Phys.* **2012**, *85*, 045440.

[17]   J. Yin, H. N. S. Krishnamoorthy, G. Adamo, A. M. Dubrovkin, Y. Chong, N. I. Zheludev, C. Soci, *NPG Asia Mater.* **2017**, *9*, 425.

[18]   B. Anasori, M. R. Lukatskaya, Y. Gogotsi, *Nature Reviews Materials* **2017**, *2*, 16098.

[19]   V. Mauchamp, M. Bugnet, E. Bellido, G. Botton, P. Moreau, D. Magne, M. Naguid, T. Cabioc'h, M. Barsoum, *Phys. Rev. B: Condens. Matter Mater. Phys.* **2014**, *89*, 235428.

[20]   J. K. El-Demellawi, S. Lopatin, J. Yin, O. F. Mohammed, H. N. Alshareef, *ACS Nano* **2018**, *12*, 8485–8493.

[21]   S. Lopatin, J. El-Demellawi, H. Alshareef, *Microsc. Microanal.* **2018**, *24*, 1578-1579.

[22]   H. Lin, X. Wang, L. Yu, Y. Chen, J. Shi, *Nano Lett.* **2017**, *17*, 384–391.

[23]   E. Satheeshkumar, T. Makaryan, A. Melikyan, H. Minassian, Y. Gogotsi, M. Yoshimura, *Sci. Rep.* **2016**, *6*, 32049.

[24]   D. B. Velusamy, J. K. El-Demellawi, A. E. El-Zohry, A. Giugni, S. Lopatin, M. N. Hedhili, A. E. Mansour, E. D. Fabrizio, O. F. Mohammed, H. N. Alshareef, *Adv. Mater.* **2019**, *31*, 1807658.

[25]   Y. Dong, S. Chertopalov, K. Maleski, B. Anasori, L. Hu, S. Bhattacharya, A. M. Rao, Y. Gogotsi, V. N. Mochalin, R. Podila, *Adv. Mater.* **2018**, *30*, 1705714.

[26]   V. K. Thakur, J. Yan, M.-F. Lin, C. Y. Zhi, D. Golberg, Y. Bando, R. Simb, P. S. Lee, *Polym. Chem.* **2012**, *3*, 962–969.

[27]   S. B. Tu, Q. Jiang, X. X. Zhang, H. N. Alshareef, *ACS Nano* **2018**, *12*, 3369-3377.

[28]   X. M. Cai, T. P. Lei, D. H. Sun, L. W. Lin, *RSC Adv.* **2017**, *7*, 15382.

[29]   M. A. Hope, A. C. Forse, K. J. Griffith, M. R. Lukatskaya, M. Ghidiu, Y. Gogotsi, C. P. Grey, *Phys. Chem. Chem. Phys.* **2016**, *18*, 5099-5102.

[30]   X.-Q. Wang, C. F. Tan, K. H. Chan, X. Lu, L. L. Zhu, S.-W. Kim, G. W. Ho, *Nat. Commun.* **2018**, *9*, 3438.

[31]   R. Y. Li, L. B. Zhang, L. Shi, P. Wang, *ACS Nano* **2017**, *11*, 3752–3759.

[32]   A. P. Straub, N. Y. Yip, S. Lin, J. Lee, M. Elimelech, *Nat. Energy* **2016**, *1*, 16090.

[33]   X.-Q. Wang, C. F. Tan, K. H. Chan, K. C. Xu, M. H. Hong, S.-W. Kim, G. W. Ho, *ACS Nano* **2017**, *11*, 10568–10574.

[34]   L. Hines, K. Petersen, G. Z. Lum, M. Sitti, *Adv. Mater.* **2017**, *29*, 1603483.

[35]   M. E. Lines, A. M. Glass. Principles and applications of ferroelectrics and related materials. Clarendon Press, London (**1979**).

[36]   M. Benghanem, *Appl. Energy* **2011**, *88*, 1427-1433.

[37]   A. K. Yadav, S. S. Chandel, *Renew. Sust. Energ. Rev.* **2013**, *23*, 503-513.

[38]   J. A. Duffie, W. A. Beckman, Solar engineering of thermal processes, John Wiley & Sons (**2013**).

[39]   R. E. Bird, C. Riordan, *J. Appl. Meteorol. Climatol.* **1986**, *5*, 87-97.

Chapter 6- Summaries and Future Work

6.1 Summaries

6.1.1 MXene-derived ferroelectric crystals: MXene-derived high aspect ratio potassium niobate ($KNbO_3$) single crystals are successfully synthesized using two-dimensional Nb_2C MXene and KOH as the potassium and niobium source, respectively. The well-defined butterfly loops of the PFM amplitude signals and the distinct 180-degree switching of the phase signals further corroborate the presence of robust ferroelectricity in M-$KNbO_3$ crystals.

6.1.2 Dielectric enhancement: MXene-containing insulating polymers evidence a large enhancement in dielectric constant. The ratio of permittivity to loss factor for MXene-percolated composites is superior to that of all previously reported fillers in the same polymer. The dielectric constant enhancement is explained on the basis of the microscopic dipole model.

6.1.3 Solar tracking system: We have demonstrated the solar tracking system based on the photo-thermo-mechanical (PTM) deformation of plasmonic $Ti_3C_2T_x$ MXene/3D aligned PVDF actuator. The sensitivity of the PTM actuator to photo stimulation is due to the marked enhancement of the surface plasmon resonance of Ti3C2Tx MXene. In this work, three types of tracking options were simulated, where uniaxial E-W options improved the overall energy intensity of solar module by over 30 % compared to the optimized tilting-control option. Our results clearly show an improvement in PV efficiency, PV power intensity, and energy intensity based on our MXene/PVDF PTM actuator. Our design provides a new way to achieve solar tracking function using photo-thermal-mechanical property of two dimensional materials.

6.2 Future Work

6.2.1 Tunable localized surface plasmonic resonance of partially oxidized 2D MXene flakes.

6.2.2 MXene self-assembled nanoparticles for electrocatalysis.

6.2.3 Up-conversion enhanced photocatalytic activity of MXene self-assembled nanowires.

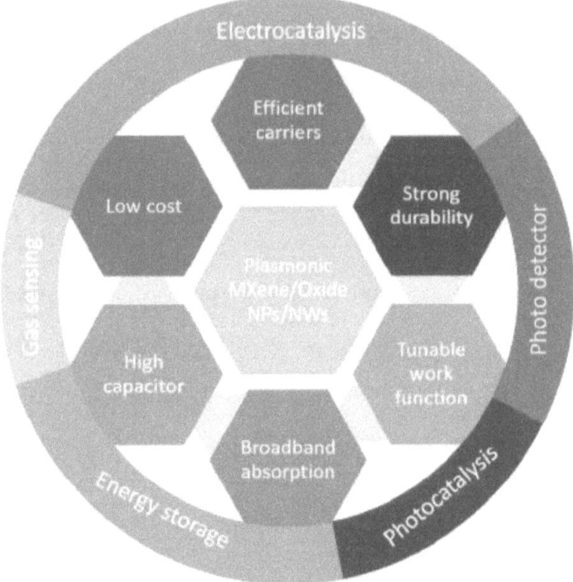

Figure 6.1 Key properties and applications of plasmon-assisted MXene based self-assembled oxides nanoparticles/nanowires.

Plasmonic MXene based self-assembled oxides nanoparticles/nanowires have promising applications in photodetector, photocatalysis, electrocatalysis, energy storage, and gas sensor.

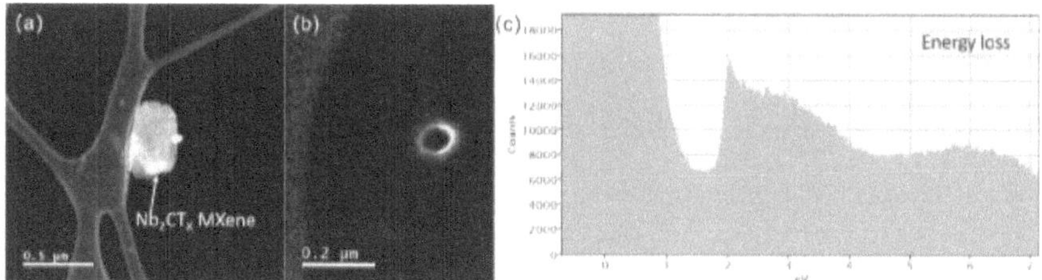

Figure 6.2 (a) TEM image of Nb_2CT_x MXene flakes on Si_3N_4 grid. (b) STEM-EELS map of Cu nanoparticles on Nb_2CT_x MXene flakes. (c) Electron energy loss spectroscopy of Cu nanoparticles on Nb_2CT_x MXene flakes.

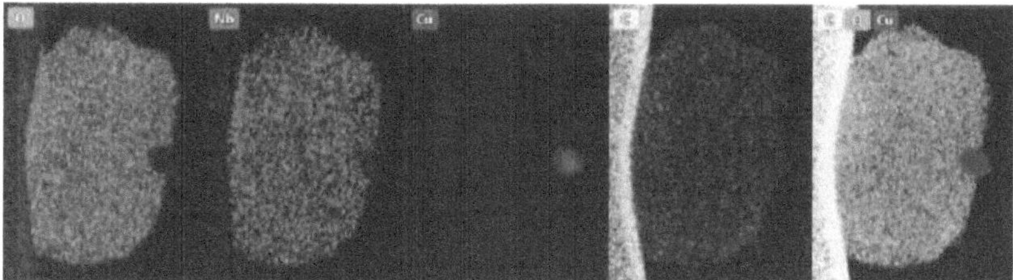

Figure 6.3 Energy dispersive spectroscopy (EDS) elemental mapping of Cu nanoparticle on Nb_2CTx MXene flakes.

Nb_2CT_x MXene has very weak surface plasmon resonance compared to Cu nanoparticles according to our recent work. NbO is metallic, NbO_2 has narrow bandgap, and Nb_2O_5 is insulating. Nb_2CT_x MXene/self-assembled NbO NPs/NWs: tunable localized surface plasmon resonance in sensing and biomedical applications.

APPENDICES

A. Fabrication of MIM Capacitors

Deposition: A film with an average thickness of 100 μm was cast on the conducting metal-coated silicon substrate, as measured by a Profilometer. Subsequently, Ti/Au contacts were deposited on the top of the MXene-polymer films through a shadow mask by e-beam evaporation to form a fully fabricated metal-insulator-metal (MIM) capacitor, which was used to measure the dielectric properties of MXene-polymer composites.

Patterning: In order to design the pattern of electrode, stainless steel shadow mask was prepared by using laser cutting. The pattern on MXene-derived crystals was fabricated by using focus ion beam microscope.

B. Characterization of MXene-derived Advanced Functional Materials

Surface Chemistry and Structure Characterization: The surface chemistry of MXene flakes were investigated by using X-ray photoelectron spectroscopy (XPS, Amicus). The structure of MXene and MXene-derived materials were studied by using XRD (Bruker D2 Phaser) and FT-IR in transmission mode (Nicolet iS10, Thermo Scientific), Raman spectroscopy (Horiba Aramis Raman Microscope), SEM (Nova Nano), and TEM (Titan Cs Image).

Electronic Properties Measurements: Frequency-dependent capacitance of MXene/polymer composite was measured with an Agilent LCR meter (4980A) in the frequency range of 1K Hz to 100 KHz and an oscillation signal of ~ 50 mVrms with a parallel equivalent circuit. The current density vs electric field (J-E) loops in our manuscript were derived from I-V curves measured by

Agilent B1500A system. The J-E loops were measured by using a direct current (DC) source with a field ramping rate of 50 V cm^{-1} s^{-1}. The dwell time was 0.5 s for each step.

Ferroelectricity Measurements: The capacitance-versus-voltage (C-V) curves of MXene-derived KNbO$_3$ crystals were measured with an Agilent LCR meter (4980A), whereas the current-versus-voltage (I-V) switching current loops were measured with an Agilent B1500A system. The hysteresis loops were measured with a commercial ferroelectric tester (Radiant Technologies, Inc. Precision Premier 200V ferroelectric test system). Piezoresponse force microscopy (PFM) measurements were carried out on the M-KNbO$_3$ crystals by applying a bias between the conductive PFM tip and the platinum-coated silicon substrate (Asylum research, MFP-3D model). The switching spectroscopic loops were recorded under resonance-enhanced PFM mode by applying an alternating current (AC) electric field superimposed on a direct current (DC) triangle saw-tooth waveform.

Photoluminescence Measurements: PL spectra of MXene-derived photoluminescent LiNbO$_3$:Pr^{3+} crystals were recorded with a fluorescence spectrometer (JASCO, FP8600), and the PL lifetime of M-LN:Pr^{3+} crystals was measured using a Tau Fluorescence lifetime spectrometer (Hamamatsu Photonics, C11367).

C. Simulation

A few simulating tools from the PV lighthouse platform were utilized to do the relevant simulation, including the solar position and spectrum simulation, light ray tracing simulation and equivalent circuit efficiency simulation. In the simulation, the albedo value was set to 0.4 under a clear sky condition. An in-house silicon heterojunction solar cell optical model plus standard EVA (conventional), glass (low Iron sodalime), and glass ARC (Magnesium fluoride) were used in the light ray racking simulation. In the equivalent circuit module, for all wavelength, the collection efficiency, ideal factor, saturation current density J_{01}, shunt resistance, series resistance, and additional grid resistance were respectively set to 0.9, 1, 10 FA/cm^2, 10 k$\Omega\cdot$cm^2, 0.8 $\Omega\cdot$cm^2 and 0.07$\Omega\cdot$cm^2.